D0708125

Illustrated Interior Carpentry

ALSO BY GRAHAM BLACKBURN

Illustrated Housebuilding

The Illustrated Encyclopedia of Woodworking Handtools, Instruments, & Devices

Illustrated Basic Carpentry

The Postage Stamp Gazetteer

Illustrated Furniture Making

An artistic representation of how early man developed building through various huts and dwellings. *from* **VITRUVIUS DEUTSCH**, *a translation of* VITRUVIUS *by* Walter Ryff, 1548

Illustrated Interior Carpentry

Graham Blackburn

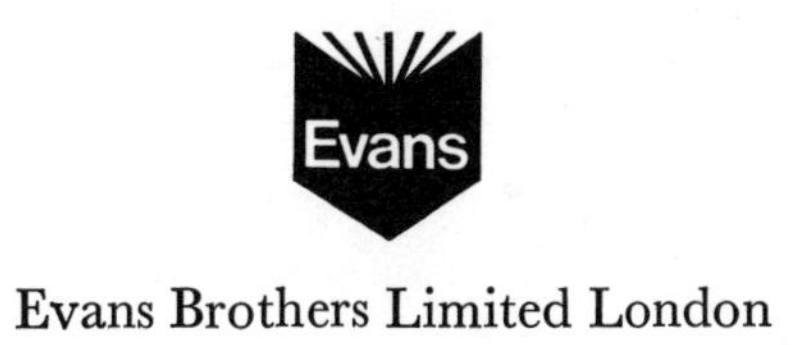

Evans Brothers Limited London

First published in Great Britain 1979 by
Evans Brothers Limited,
Montague House, Russell Square,
London, WC1B 5BX

Original edition first published by
The Bobbs Merrill Company Inc.,
Indianapolis · New York

Designed by Graham Blackburn

Printed in Great Britain by
Hazell Watson & Viney Ltd, Aylesbury, Bucks
ISBN 0 237 44973 0 PRA 6195

for Heide Duggal

SPECIAL THANKS TO

My favorite editor, Evelyn Gendel, who helped conceive this book; my favorite publisher, Peter Mayer, who gave me my start and thus made all my books possible; my best friends, Barry Cahn, John Holbrook, and Paul Potash, who at various times looked after me during the writing of this book; Lynda Redfield, Denis Dehr, and Tom Laine, who in different ways provided the high points of the year; Nick Jameson for being so generous with his Steinway; Māra Lepmanis for being so special; Elaine Louie, Thomas Holman, Douglas Holman, and Geoffrey Stephenson for making life fun; Sandra Shaw for being Maid Marian all the way from Woodstock to Houston and Appaloosa all the way from Houston to Los Angeles; Carol Glenn for being my favorite photographer; Thomas Gayton for enlivening Neptune Beach; Carolyn, Rebecca, and Sabrina Berliner for being my favorite family in Hollywood; and Tina Whitney Firestone, without whom my (and many other people's) life would be much duller.

Contents

Introduction

Encouraged by the success with which my earlier book ILLUSTRATED HOUSEBUILDING (Overlook Press, 1974) has met, I have written this book to follow on and, to a certain degree, to complete many of the topics necessarily given only scant attention in the earlier work.

The present book is, however, organized along different lines and need not be read from beginning to end, but may be dipped into here and there as the need dictates. There is a certain logic to the order in which the various subjects are treated, but so many things must be done in overlapping stages that it would be impossible to describe coherently the finishing of a house in strict chronological order, and the various subjects are therefore treated separately, one at a time.

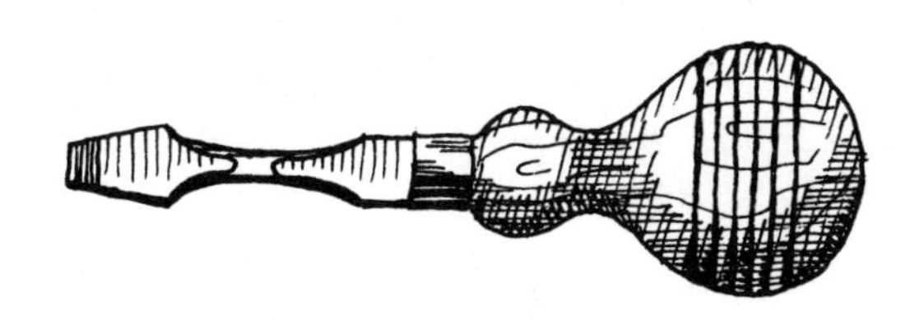

It should be borne in mind that the main purpose of this book, like that of its predecessor **ILLUSTRATED HOUSEBUILDING**, is to enable the average person to build and complete a house. It is not, therefore, an exhaustive treatise on all aspects of wood-working and joinery. Many things are left out because they are beyond the scope of what is required for a small house. However, for those who would know more, there is a bibliography wherein will be found access to more detailed information.

The main requisite has been that there be enough information for the layman to build well and with a choice of alternatives. Those who would become professional carpenters, joiners, or architects must certainly know what is written here, but they will also have to know much more if they are to become fully competent in their fields.

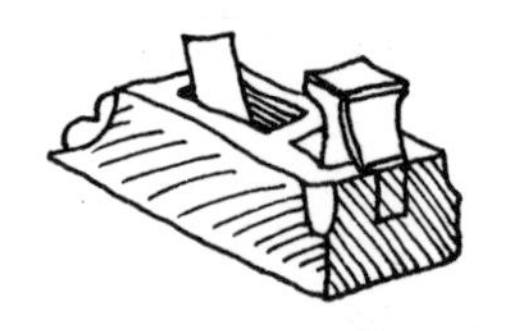

Nevertheless, I would like to emphasize that I have in no way pandered to the weekend-dabbler or half-interested do-it-yourselfer. The information presented here is, I hope, completely "professional." If anything, I have erred on the side of craftsmanship rather than on the side of expediency. Indeed, many shoddy tricks and short cuts are to be met within so-called "professionally built" houses today. It is a shame that with so many new and wonderful materials at our disposal – such as thermopane, fiberglass insulation, Formica, and even sheetrock and plywood – that so little "craftsmanship" is in evidence in new buildings. There are, of course, exceptions, but so much of the work covered in this book, namely the millwork, finish carpentry, and joinery, that goes on in even high-cost dwellings is shoddy in extreme compared to the same work of even fifty years ago.

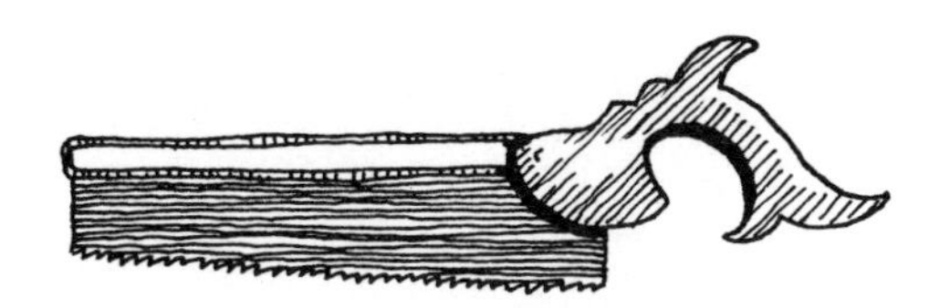

It is not only the cost – in terms of money and time – of our shelter that is important, but also the beauty and craftsmanship with which we surround ourselves. For these things, despite the press of financial considerations, are ultimately more important.

And for the reader of this book there is something else that is important: a door, for example, though not mechanically perfect, is worth infinitely more if we have, with intelligence and application, made it ourselves rather than bought a possibly more expensive installation from a "home components fabricator."

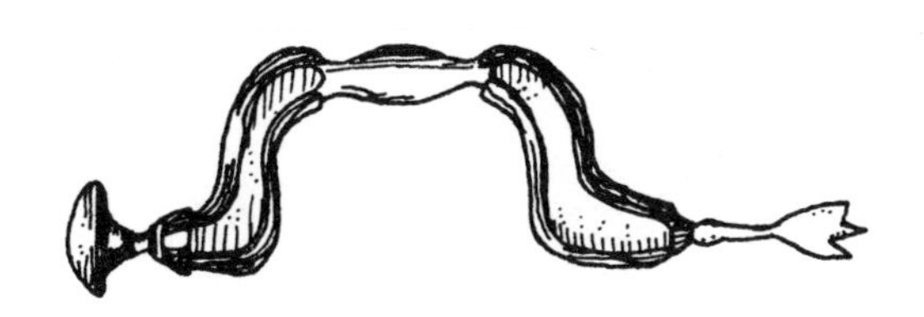

Finally, it is assumed that the reader will have a passing acquaintance with basic woodworking skills. However, the most important requirements are integrity and patience, for, though skill and technique may come with practice and experience, little of value will ever be achieved if we do not care.

Graham Blackburn
LA JOLLA, CALIFORNIA, 1977

AUTHOR'S NOTE

Dimensions

The subject of dimensions is at the moment somewhat confusing, firstly, because of the system of classifying milled lumber in terms of sawmill cuts (that is to say, although the sawmill may cut an exact 2" × 4", by the time the wood has been seasoned, dried, and milled, it may measure anything from $1\frac{1}{2}$" × $3\frac{1}{2}$" to $1\frac{11}{16}$" × $3\frac{11}{16}$" and yet still be called a "2" × 4"); and secondly, because of the changeover to the metric system, which is imminent in America and already under way in Britain.

Ultimately, the adoption of the metric system will have simplified matters greatly; indeed, in Britain new standards are being laid down which are a vast improvement over the old "imperial" system of inches and feet. Therefore, I have avoided the use of actual measurements where not strictly necessary, but where they occur, bear in mind that lumber measurements are "nominal" and other measurements are "standard" American.

In general, it is the proportions which are important, and although a millimeter may be more convenient than an inch, in the long run it will matter little what the proportional units were called at the time of construction.

Carpentry & Joinery

CARPENTER : a workman who shapes and assembles <u>structural</u> woodwork, especially in the construction of buildings.

JOINER : one who does the woodwork (as doors or stairs) necessary for the <u>finishing</u> of buildings.

WEBSTER'S THIRD NEW INTERNATIONAL DICTIONARY

CARPENTER : one who does the framework of houses. . . as opposed to a joiner.

JOINER : a worker in wood who does lighter and more ornamental work than that of a carpenter.

THE OXFORD UNIVERSITY DICTIONARY

According to the dictionary definitions, this book should be properly entitled "Illustrated Joinery"; however, I have chosen to call it "Illustrated Interior Carpentry" for the following reasons:

Whereas the dictionary distinction between "carpenter" and "joiner" was commonly observed in years past, there are few people now who would call for a "joiner" when needing "joinery" work; rather would they call for a "carpenter" and think of the work discussed in this book as "carpentry" — or at most "finish" or "trim" carpentry." The "carpenter" of yesteryear is now referred to, in America at least, as a "framer."

Although some may object to this apparent semantic inconsistency, it should be pointed out that regardless of dictionaries or pedants, it is common usage which is the ultimate arbiter in questions of meanings. We must accept changes of meanings as part of the natural growth of language – after all, as many dictionaries will evidence, many, many words began with meanings quite different from their present ones, including "carpentry," which originally meant "carriage-making"!

Joinery is a branch of Civil Architecture, and consists of the art of Framing or joining together wood for internal and external finishings of houses . . . Hence joinery requires much more accurate and nice workmanship than carpentry . . .

Peter Nicholson 1812

Chapter One

Doors

Most doors may be classified as either exterior doors or interior doors. Exterior doors include front entrance doors, back entrance doors, and French windows. Interior doors include doors between rooms and closet doors.

Aside from the fact that exterior doors are usually larger and thicker than interior doors, both types share a variety of designs. Some of these are listed below.

1. LEDGED AND BRACED DOORS

Sometimes known as "batten doors" or "country-style doors," these consist of vertical boards, usually tongued and grooved together, with ledges and braces on the reverse side. Some typical examples are illustrated on the next page.

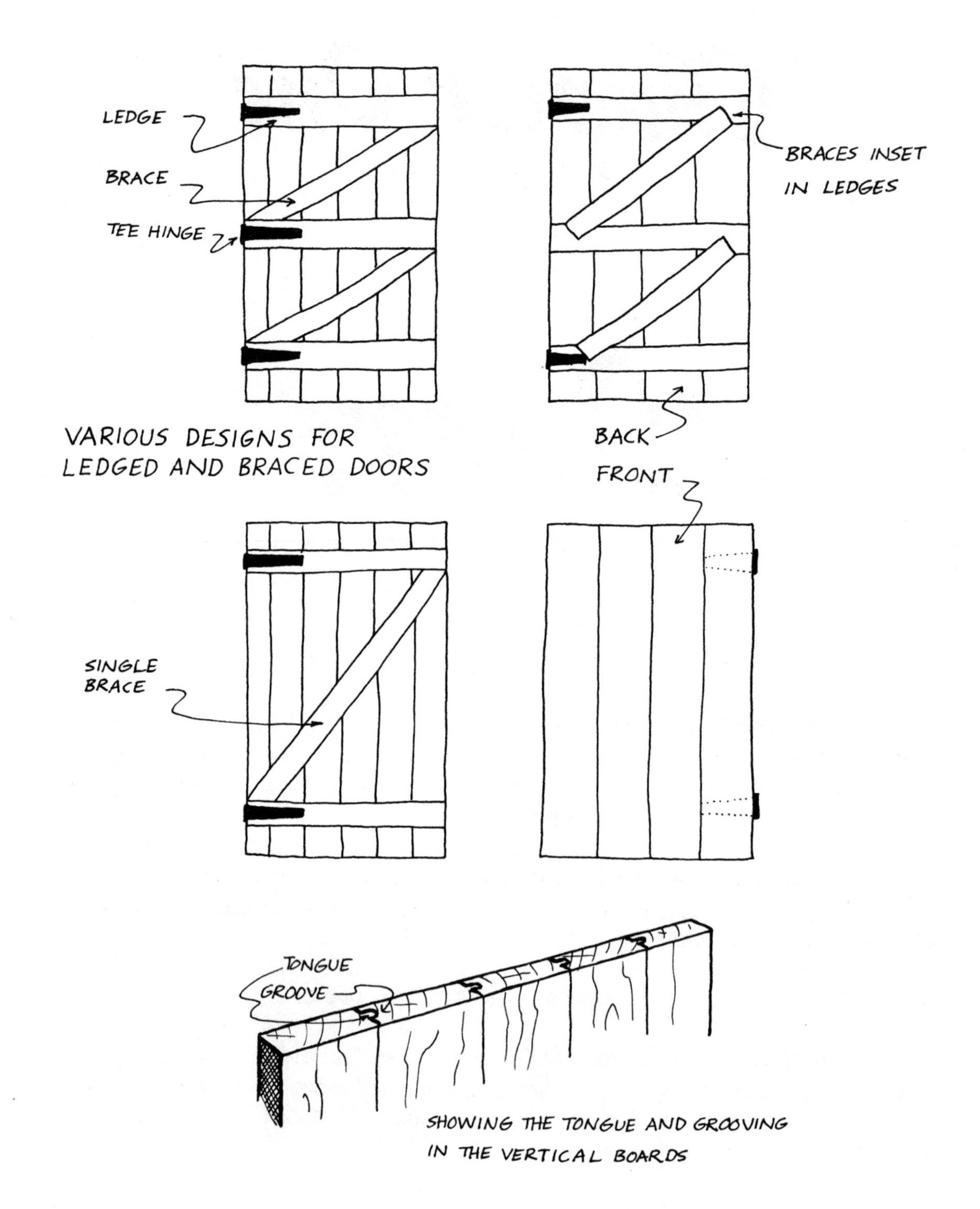

VARIOUS DESIGNS FOR LEDGED AND BRACED DOORS

SHOWING THE TONGUE AND GROOVING IN THE VERTICAL BOARDS

2. FRAMED LEDGED AND BRACED DOORS

Framed ledged and braced doors are a more sophisticated version of ledged and braced doors. They consist of a mortised and tenoned outer frame with a sheeting of tongue and groove matched boards, ledged and braced as above. "Matched" means that the ends, as well as the sides, are tongued and grooved.

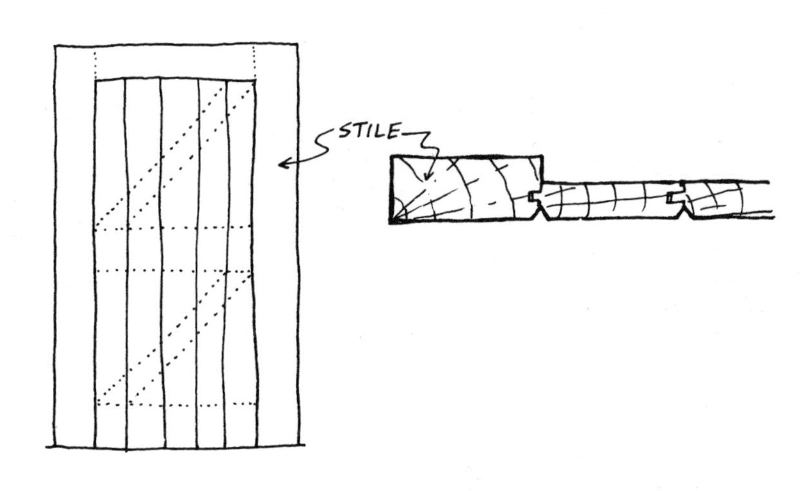

3. BEAD AND BUTT PANELED DOORS

Bead and butt paneled doors are flush panel doors, more commonly used in Britain than in America for rear entrances. The panels are tongued into the framing, being first beaded on their vertical edges.

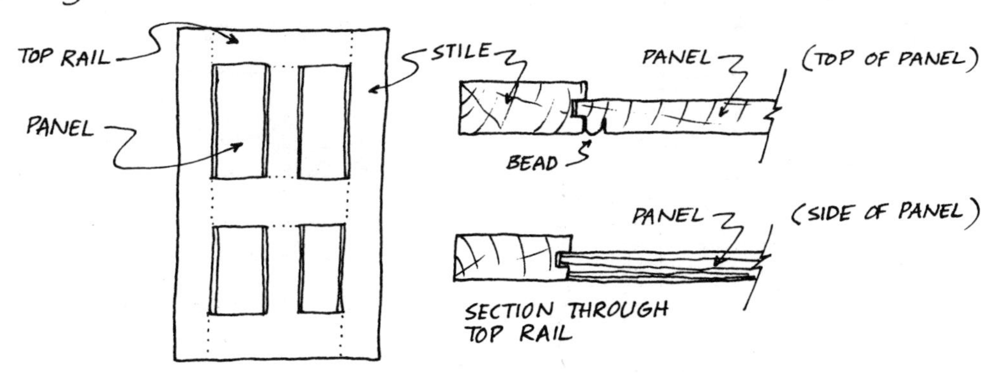

4. PANELED DOORS

Paneled doors consist of a frame made of stiles, rails, and muntins, with panels grooved into the framing. Exterior paneled doors very often have the top, or frieze panels, made out of glass. The number of panels is variable and ways of fixing the panels are numerous.

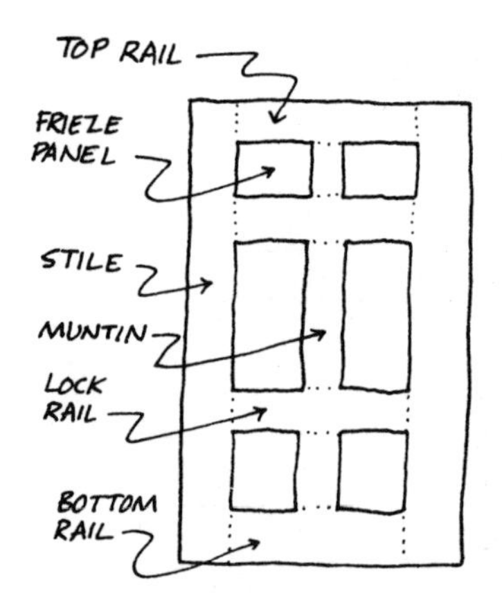

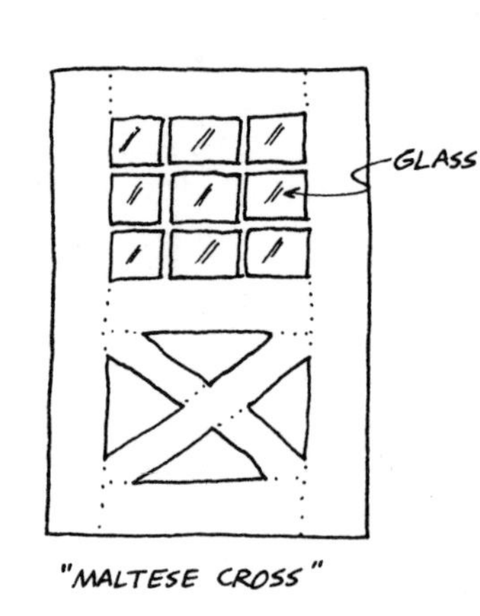

"MALTESE CROSS"

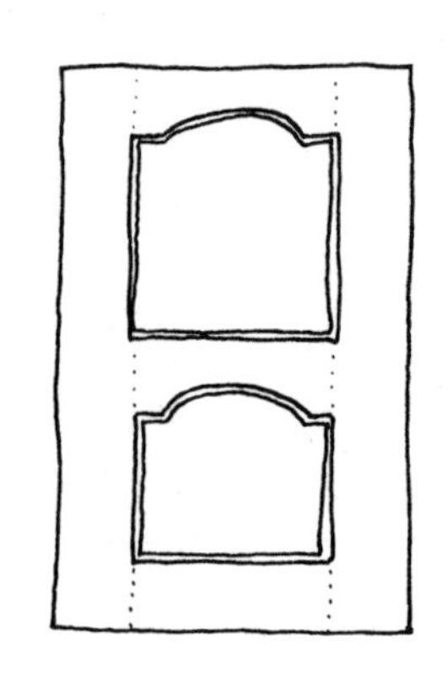

VARIETIES OF PANELING

TRADITIONAL RAISED PANEL DOOR

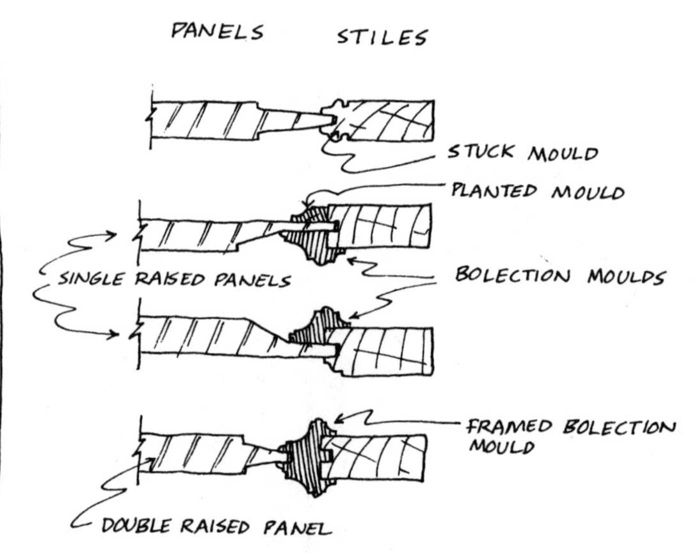

DIFFERENT WAYS OF FIXING PANELS

5. GLAZED DOORS

Glazed doors may be fully glazed, such as French windows, or half glazed. The door may be glazed with putty or glazing compound, or fitted with wooden beads. (Sliding glass doors are discussed in Chapter Two.)

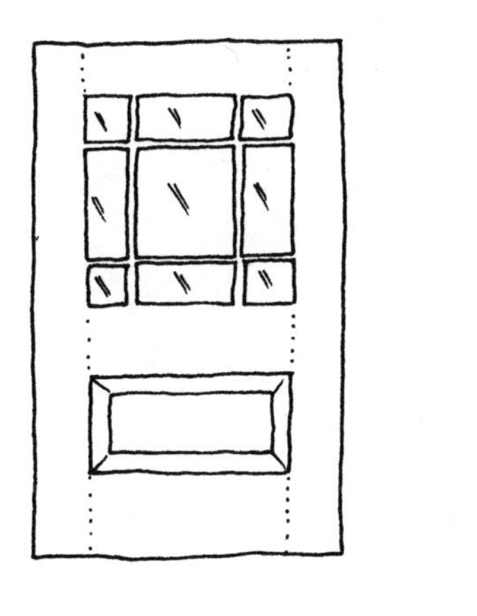

HALF GLAZED

FULLY GLAZED

6. INTERIOR PANELED DOORS

Interior paneled doors are usually thinner and narrower than exterior paneled doors. Two typical interior designs are shown below.

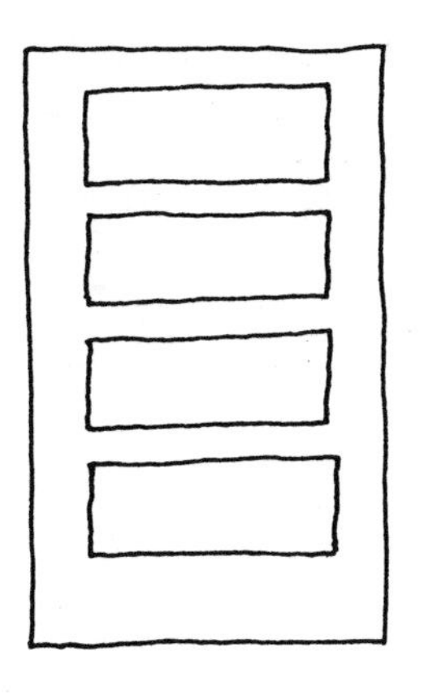

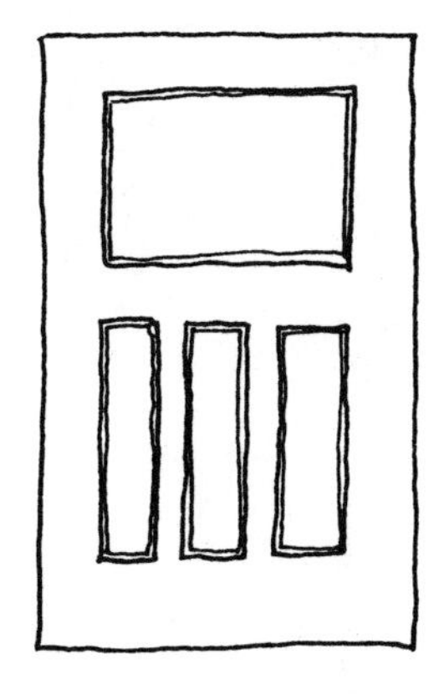

7. FLUSH DOORS

Flush doors consist of thin plywood faces covering a wooden frame with a solid or a hollow core. The plywood faces are made from a variety of woods, including birch, oak, gum, and mahogany. Even though they are called "flush," these doors are sometimes decorated with false panels and applied mouldings.

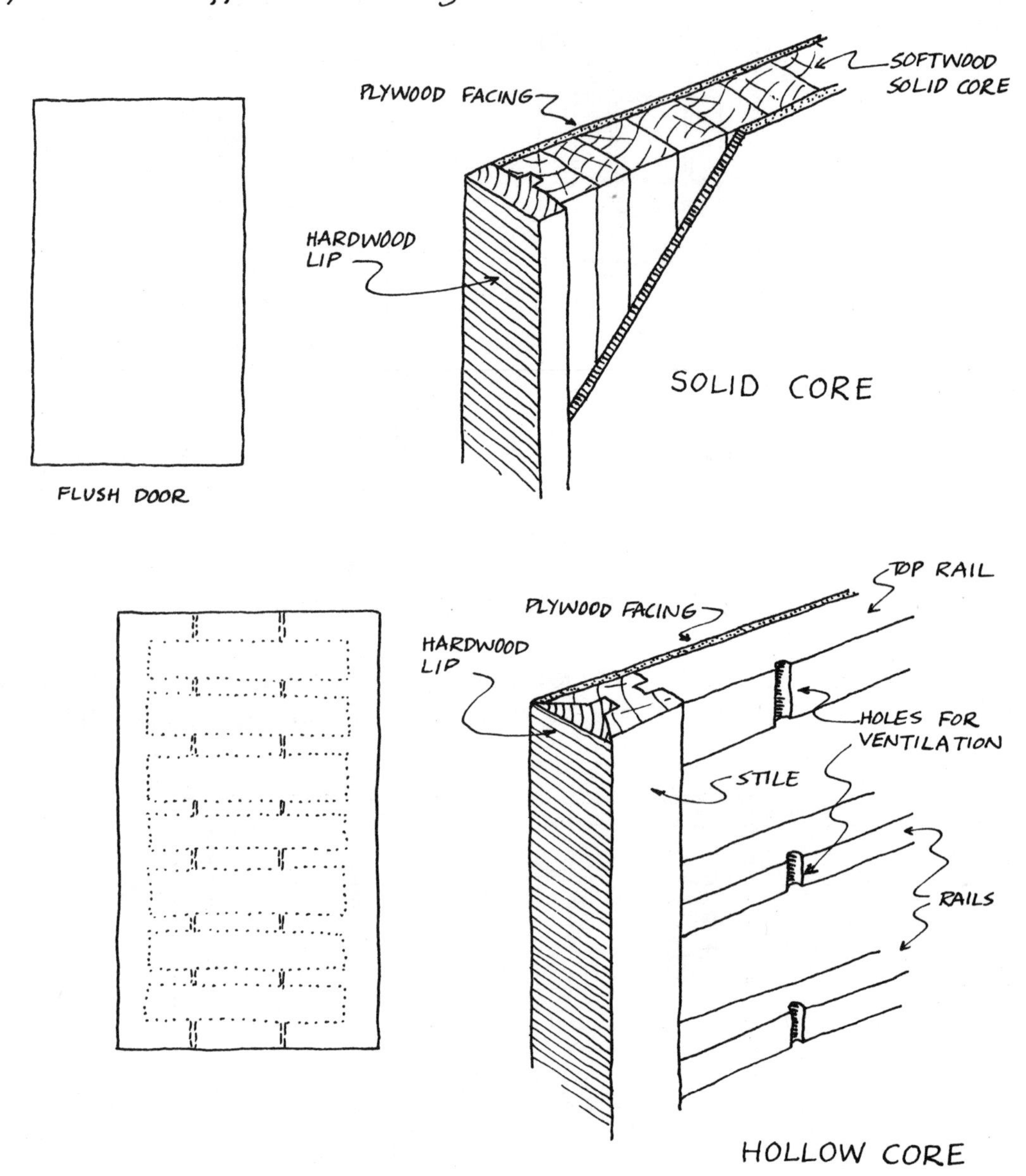

8. PREHUNG THERMAL DOORS

Prehung doors are complete, factory-made units, prehung and often prefinished in their own frames. They are relatively easy to install, eliminating much of the traditional carpentry associated with hanging and installing doors. Unfortunately, many of the designs are imitative of older styles and the fake panels and applied mouldings give these units a strangely insubstantial look. Their main advantage lies in the fact that they often use insulating glass and "thermal exterior cores" and have been precision-engineered to eliminate drafts, weather, and dust infiltration.

9. MISCELLANEOUS DOORS

By this group I mean to include sliding doors, folding doors, revolving doors, louvered doors, and other novelty doors. Most of these, however, are types by virtue of their fitting rather than their intrinsic design.

SOUTH DOOR OF ST. HELEN'S CHURCH
BISHOPSGATE · LONDON

CONSTRUCTION

1. LEDGED AND BRACED DOORS

Doors of this kind are sometimes made without the diagonal pieces, called braces (which are, however, in reality not "braces" but "struts," designed to transmit the weight of the door to the inner ends of the middle and bottom ledges, and so prevent the dropping of the "nose" or free end of the door), and, as such, are properly known as "ledged doors," but this practice is inadvisable, since the braces also help greatly in preventing the door from warping, a very common fault in plain ledged doors.

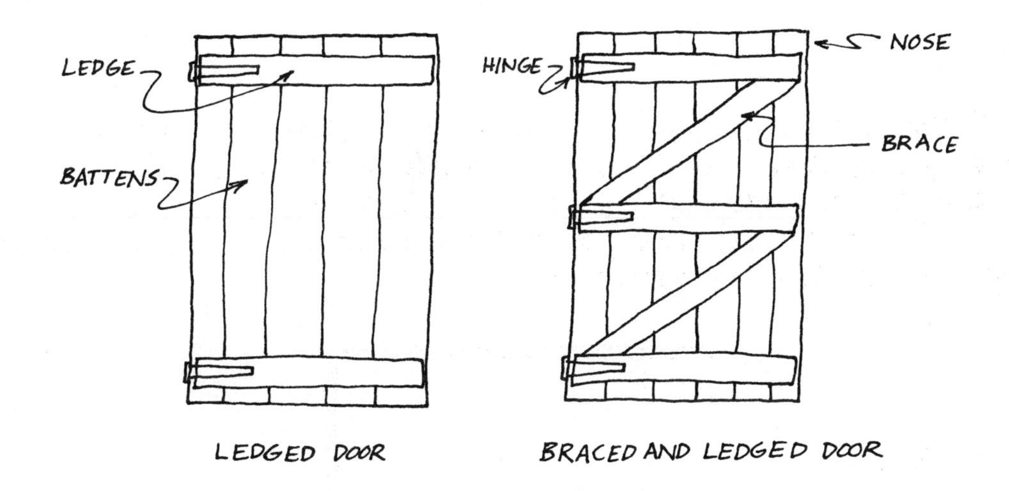

LEDGED DOOR

BRACED AND LEDGED DOOR

The size of the lumber used and the finish size of the door will depend upon what is available, your own preferences, and the size of the opening into which the door is to fit. Suffice it here to point out that doors for outbuildings need to be stouter than interior doors, and while 1" stuff* is sufficient for the latter, you might use 2" stuff† for the former.

* 25 mm – this is the approximate metric equivalent, not yet a standard.
† 50 mm – ditto.

The first step is to butt together the edges of the battens (these are the vertical pieces), which may be square-edged boards, but preferably will be tongued and grooved (sometimes referred to simply as T and G).

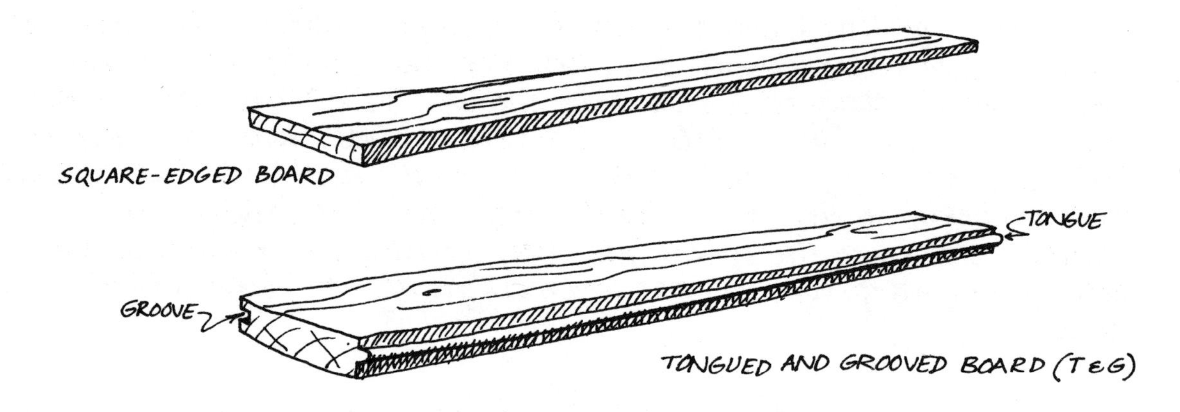

The battens should be a little longer than the required height of the door, and, when tightly together, a little wider than the finish width of the door. Thus all edges may be sawn square at one time, and the tongue on one side and the groove on the other be removed.

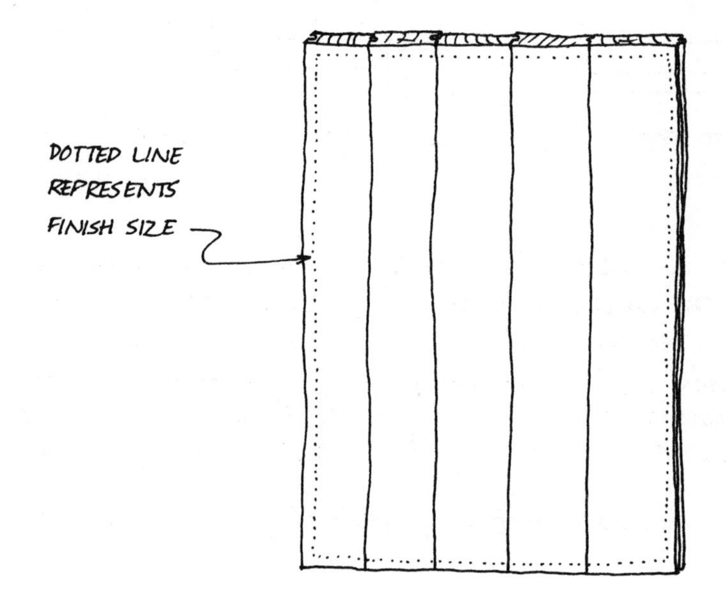

Unless you are using really well-dried wood, it is liable to shrink somewhat, so it is important that, initially at least, the battens be fitted tightly together. In good class work the edges are painted; this makes the door more weatherproof. The battens are held together by the ledges, which may be made of the same tongued and grooved material - with the edges sawn off as below.

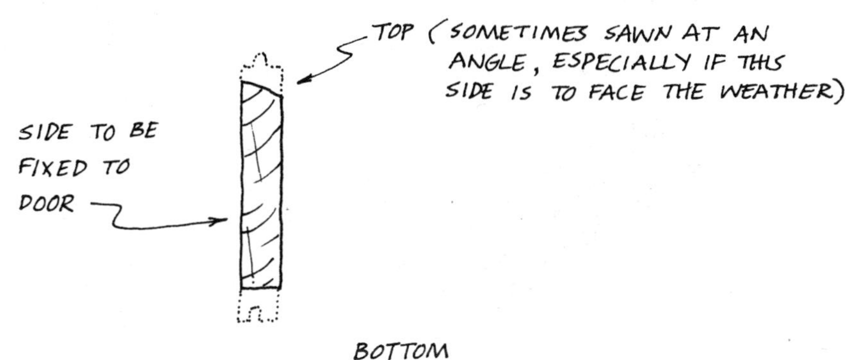

The ledges are screwed into the outside battens and nailed to the inside battens. To hold the battens tightly together while the ledges are being fixed, you may either nail blocks to the floor and insert wedges, as shown, or simply use bar or pipe clamps.

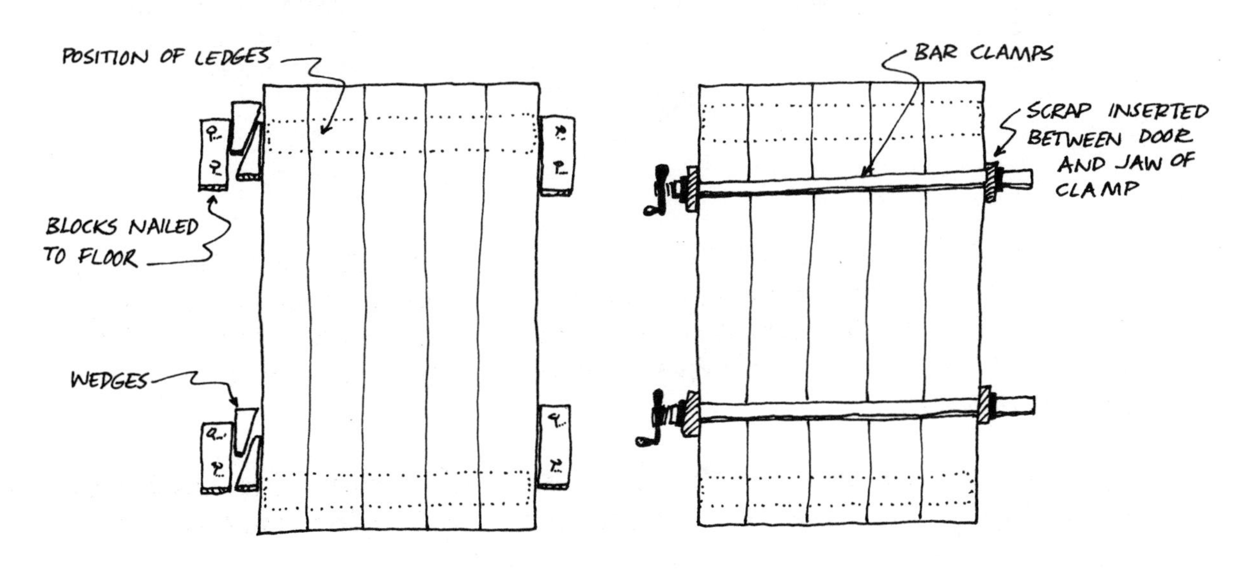

The method of screwing and nailing the ledges is illustrated below.

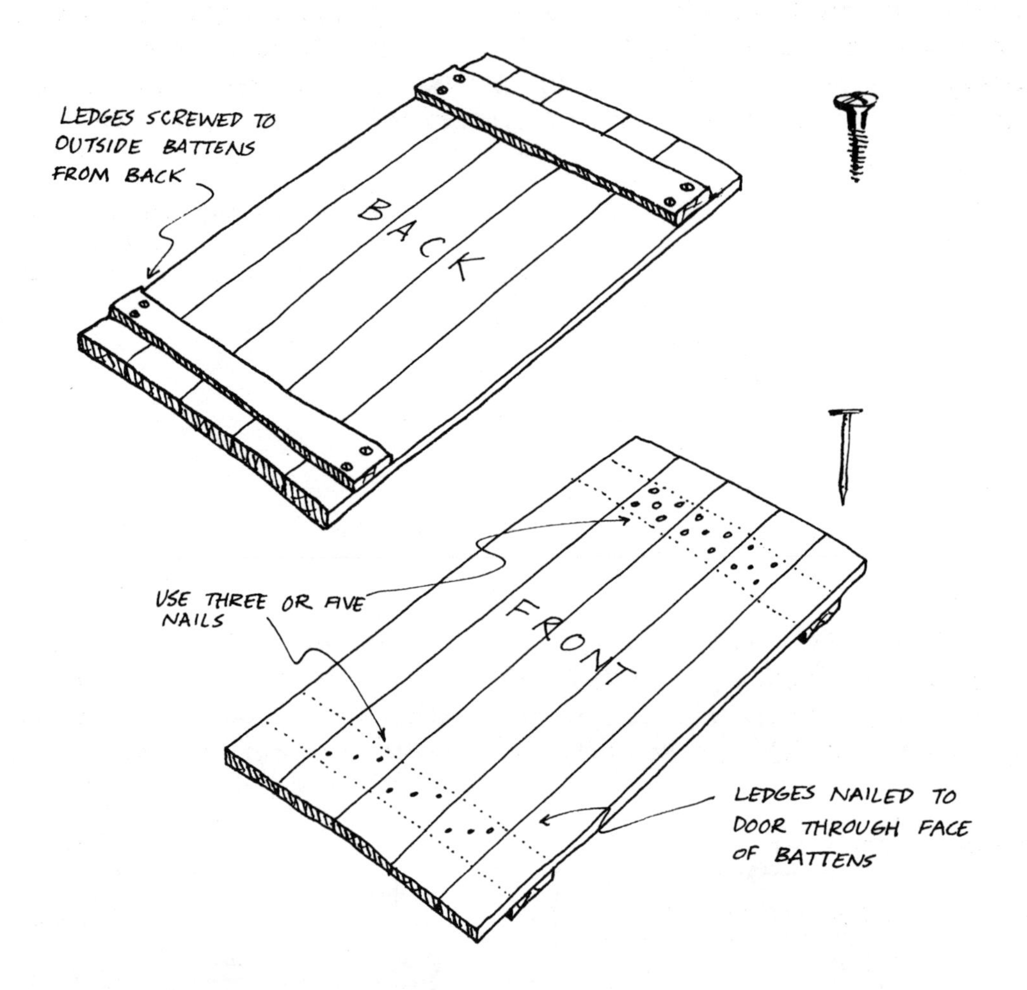

Note that the end battens, held by screws to the ledges (to prevent the door from sagging – as it might do if only nails were used), are screwed through from the back – the ledged side of the door – while the nails are driven through from the front, or face, side of the door.

The nails used should be $\frac{1}{4}$" or so longer than the combined thickness of the ledge and the battens, so that after having been driven through, the ends may be clinched over, using a nail set.

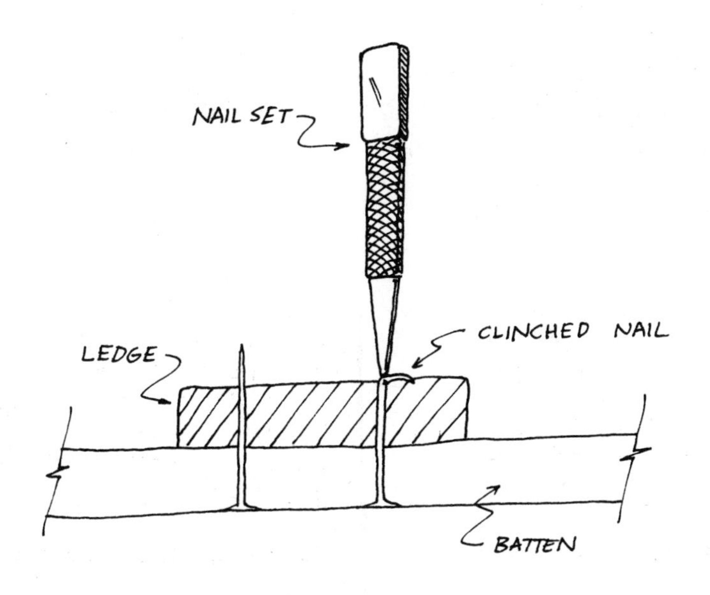

Depending on how the door is to fit in its frame – which will be discussed later under the heading **INSTALLATION** – the ledges may or may not extend the full width of the door. This is readily understood from the illustration below.

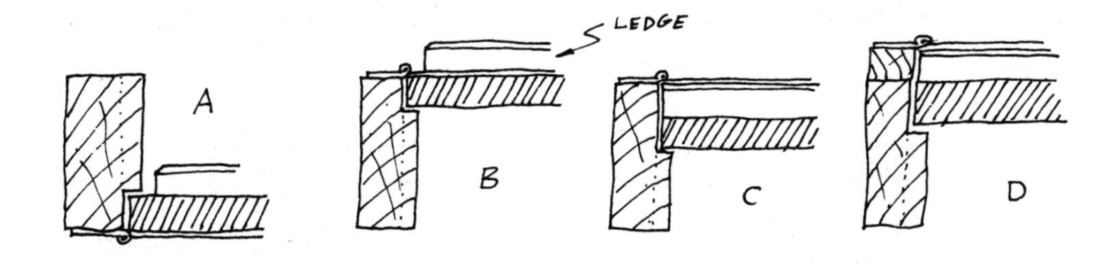

A DOOR OPENING OUTWARDS, HINGES FIXED TO FACE OF DOOR
B DOOR OPENING INWARDS, HINGES FIXED TO BACK OF BATTENS
C DOOR OPENING INWARDS, HINGES FIXED TO LEDGES
D DOOR OPENING INWARDS, HINGES FIXED TO LEDGES, FRAME RABBETED TO DEPTH OF BATTENS, THEN BLOCKED OUT

The ledges should not be more than 3" or so from the top and the bottom of the finished door, or the ends of the battens may suffer damage when the door is in use.

Furthermore, if the braces are to be notched into the ledges rather than simply butted, the notches must be cut in the ledges before they are fixed to the door! So bear all this in mind before beginning to nail.

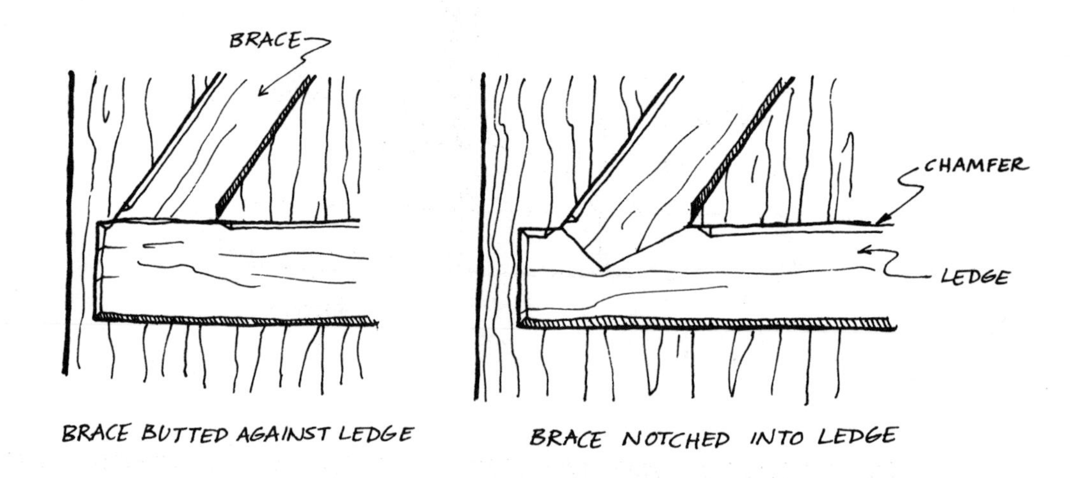

BRACE BUTTED AGAINST LEDGE

BRACE NOTCHED INTO LEDGE

Exactly where the brace meets the ledge is a matter of design, as is whether you chamfer the edges, as in the illustration above, but remember that the braces only work if they run downwards toward the hinged side.

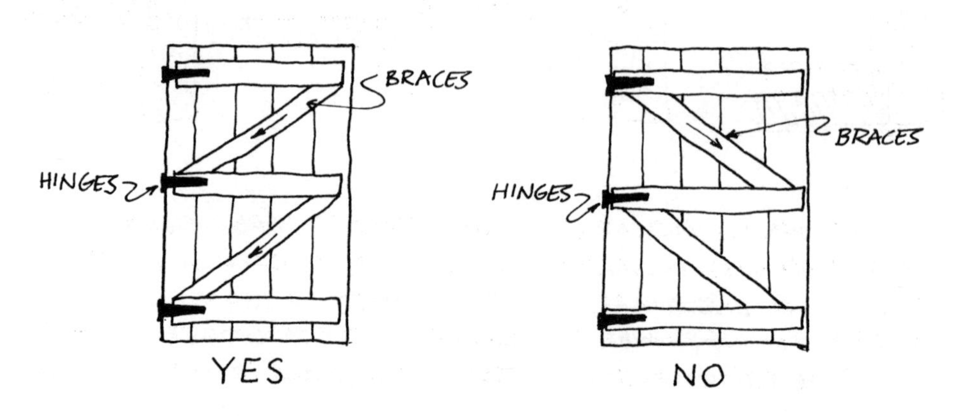

YES

NO

The procedure for notching is as follows: First cut the ledges to shape and fix them to the battens. The actual shape of the notch may vary considerably, as shown below.

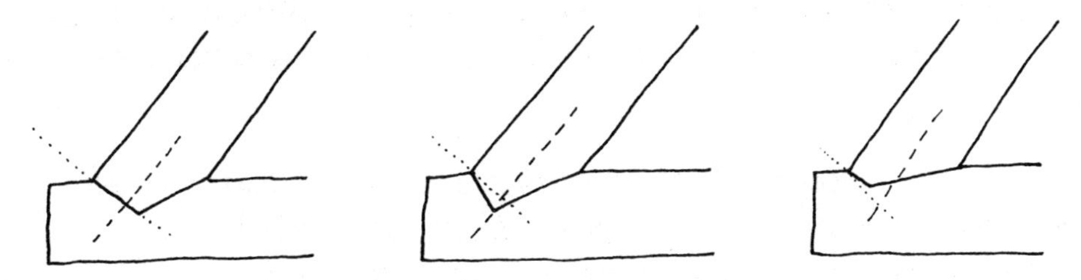

DOTTED LINE IS AT 90° TO THE BRACE
DASHED LINE RUNS DOWN CENTER OF BRACE

When the ledges are notched and fixed to the battens, mark the ledges as at A, below. Then lay the brace over the ledge, as at B, below. Now mark the brace as at C and D, below.

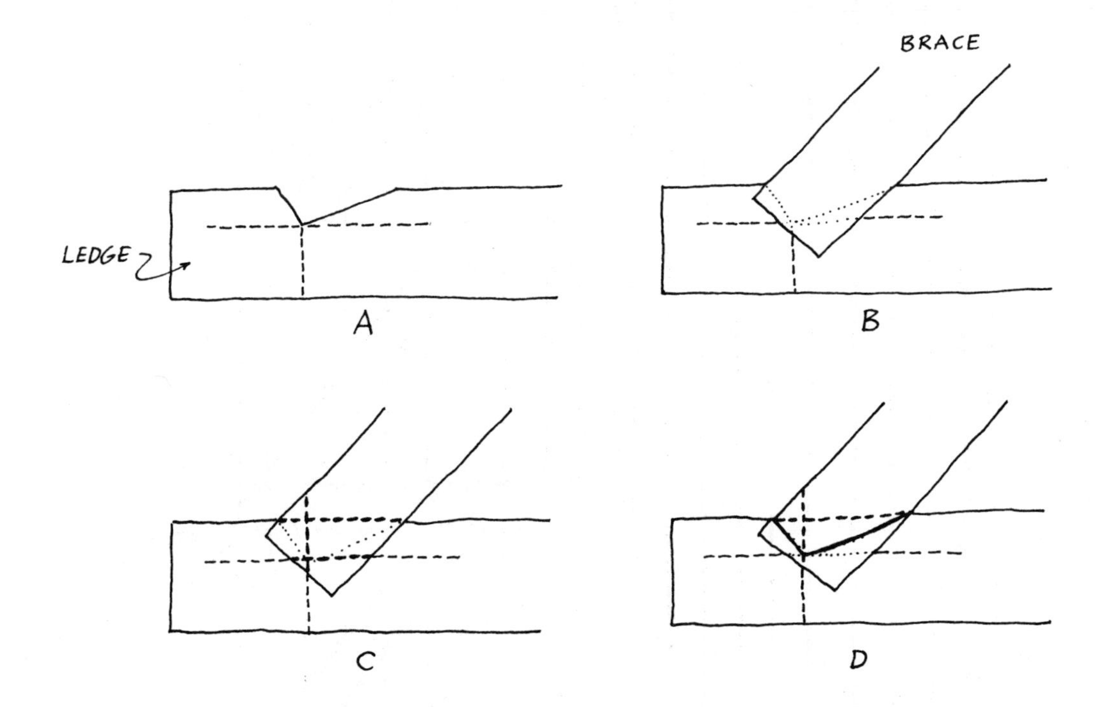

When marking the ends of the braces - whether notched or simply butted - mark both ends at the same time, lightly tacking the brace to the ledges if necessary to ensure that it doesn't slip.

After having cut the braces, nail them in place. The door is now (apart from trimming to size, which is done when it is hung, along with being fitted with its various hardware) complete.

Before the days of electrically operated, horizontally hung garage doors, ledged and braced doors were also much used for situations where doors larger than normal interior doors were needed, such as the stable doors illustrated below.

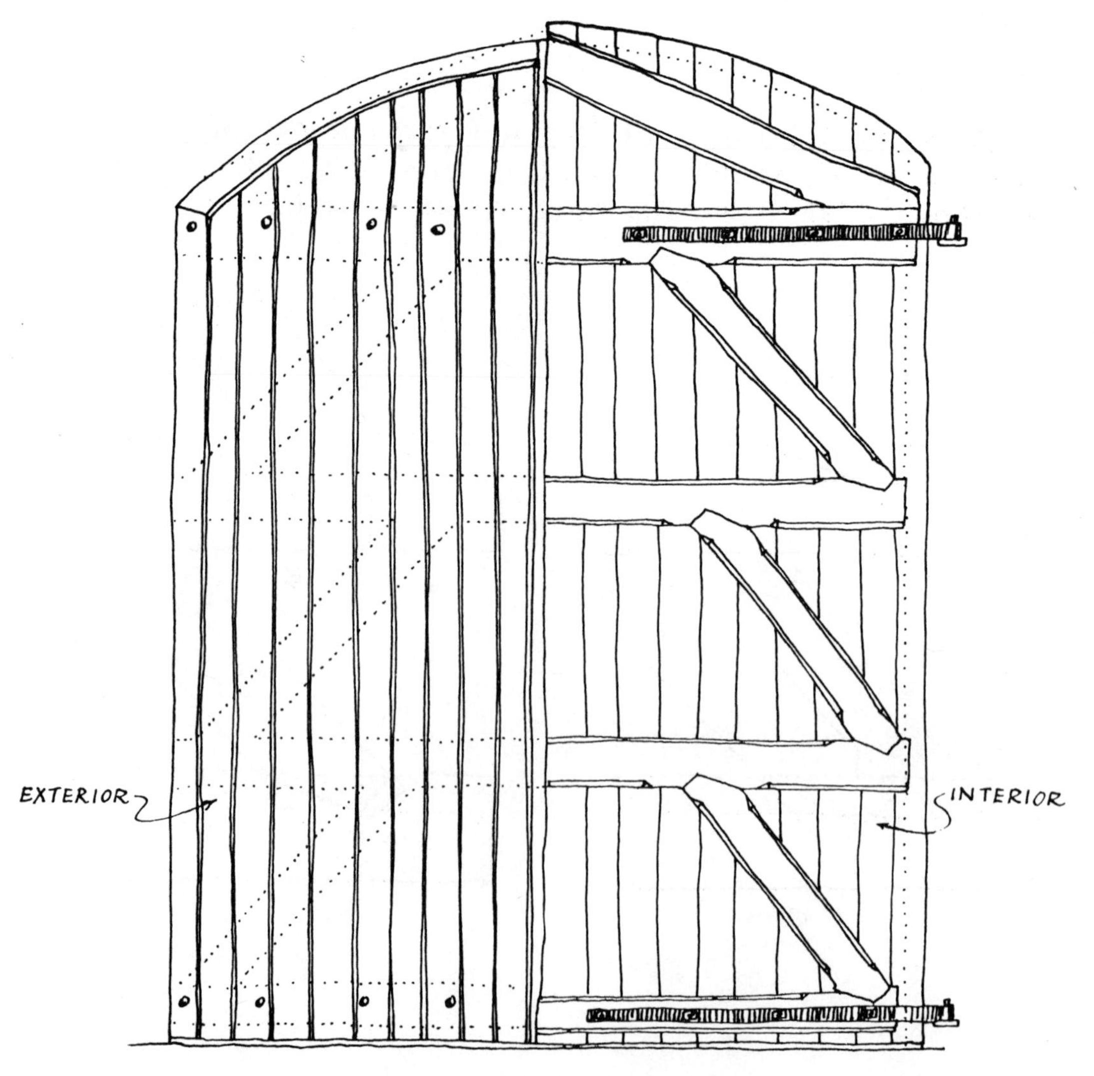

2. PANELED DOORS

Together with flush doors, whose construction will not be discussed, since they are more suited to factory manufacture than being made "on the site," paneled doors comprise the most frequently used type of interior and exterior doors.

The variations in design are almost endless. They may be plain, chamfered, beaded, or moulded on one or both sides, and the panels themselves may be worked separately and fixed on with wooden pegs or nails (technically known as "planted"), or worked on the solid (known as "stuck"). Needless to say, the overall shape of the door may vary too; the heads of the doors may be square or arched, and the panels themselves may not always be rectangular.

DOUBLE ENTRANCE DOORS WITH ELLIPTICAL FANLIGHT AND ROUND CENTER PANEL

In the framing of doors, as in other framing, the vertical pieces are known as "stiles" and the horizontal pieces "rails." Any intermediate vertical piece which is tenoned into mortises formed in the rails is called a muntin (from the French "montant," meaning "mounting"). The rail next above the bottom rail is called the "lock rail." If there is another rail between the lock rail and the top rail, it is known as the "frieze rail." Similarly, the various panels are known as "frieze panels," "middle panels," and "bottom panels."

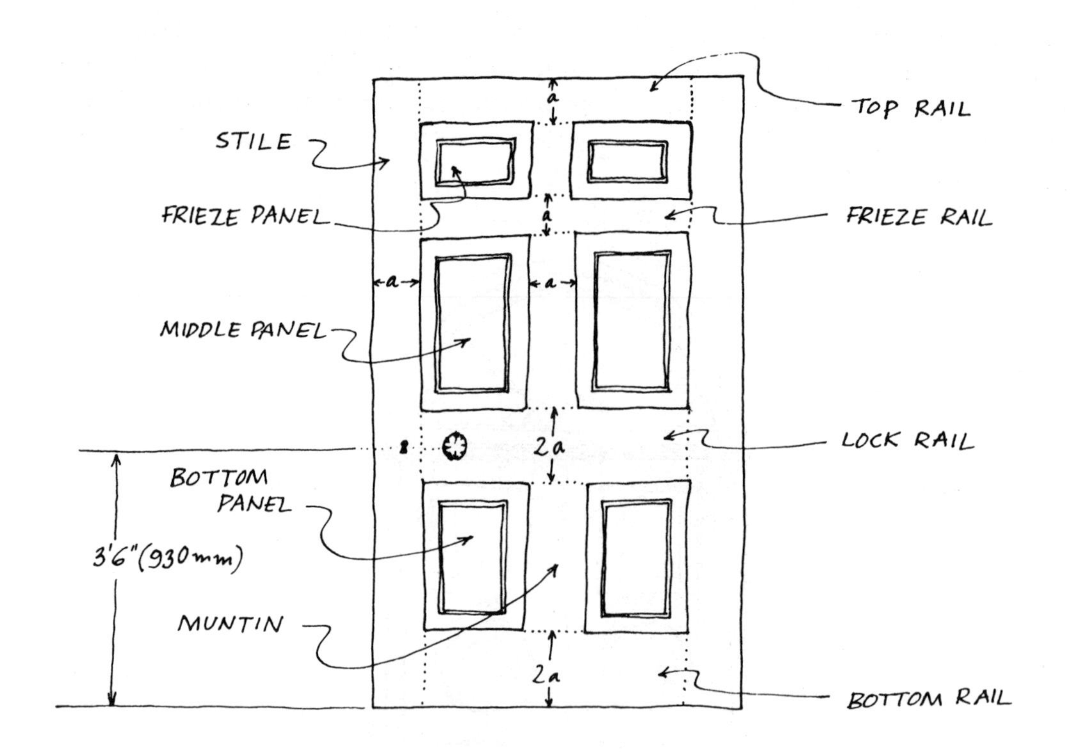

In the majority of framed doors, the top (and frieze) rails are usually the same width as the stiles ("a" in the illustration above); the bottom and lock rails usually twice as wide ("2a" above).

If the lock, or handle, is fixed at the level of the middle of the lock rail, this is usually about **3' 6" (930mm)** from the bottom of the door. If the lock is fixed entirely within the stile then the lock rail may be higher or lower, as fancy dictates.

The most ordinary form of paneled door is that with four oblong panels such as illustrated below. Being devoid of any moulds or beading, it is known as a "plain" or "square-framed" door.

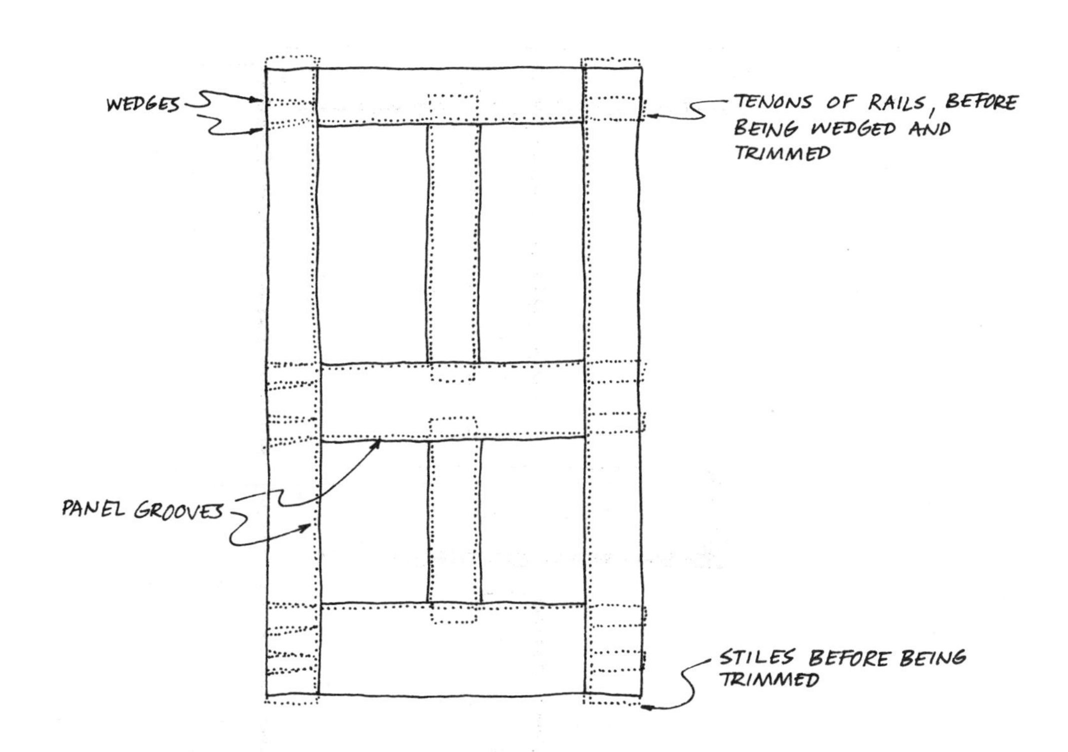

The stiles should be cut a little longer than the height of the door and the rails somewhat longer than the width, so that the door may be uniformly trimmed after assembly (as with the ledged and braced door) and also, of course, to ensure that the mortises in the stiles are completely filled.

Each muntin must be about 4" (100 mm) longer than the adjacent panels, in order to allow for a tenon, 2" (50 mm), at each end.

Take care when making up the framework that it be of uniform thickness and finished true; that is, that each piece be perfectly square to all its edges.

The stiles and top and bottom rails must be plowed along one edge to receive the panels, and the lock rail and muntins along both edges.

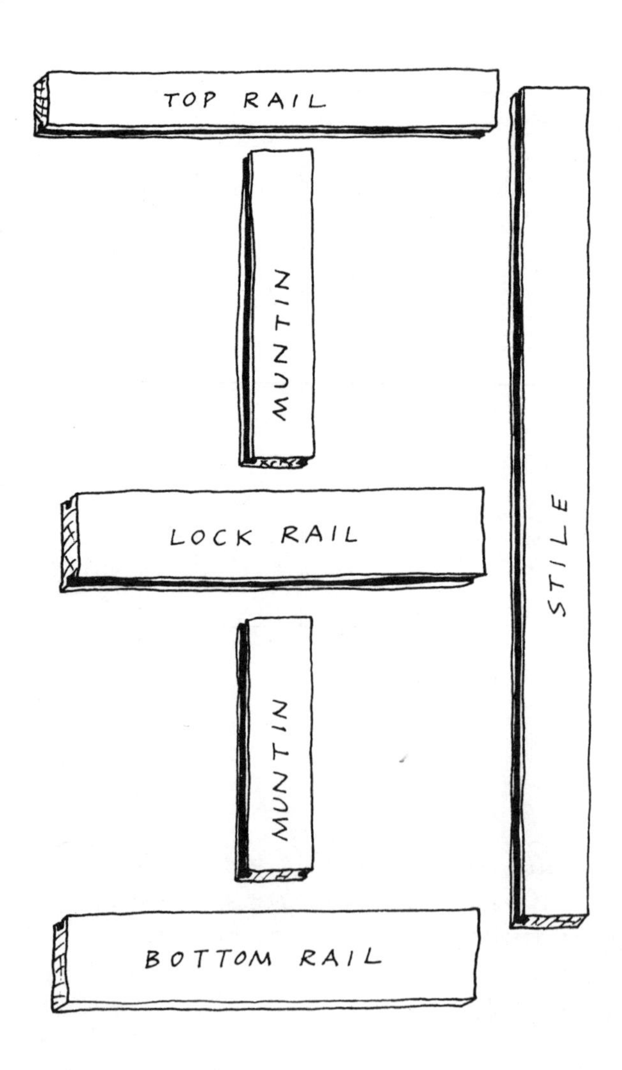

(LEFT SIDE STILE NOT SHOWN FOR CLARITY'S SAKE)

The bottom of the groove in the top rail forms the bottom of the tenons at the ends, as shown below.

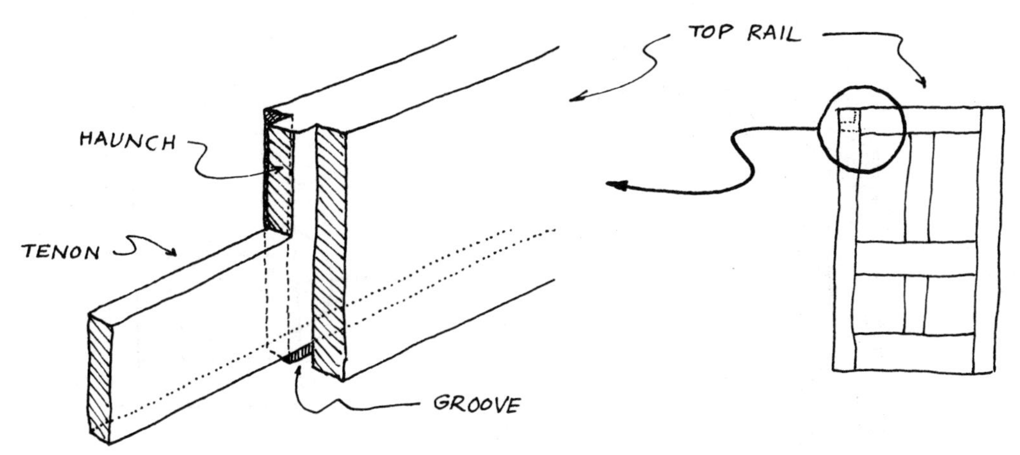

NOTE : HAUNCH IS $\frac{1}{3}$ THE WIDTH OF RAIL
TENON IS $\frac{1}{2}$ THE DEPTH OF RAIL

Therefore, the depth of these tenons should not be much more than half the depth of the rail, for otherwise the part of the stile above the mortise may be sheared out when "wedging-up."

A haunch must be left above the tenon to fill the panel groove of the stile.

The width of the tenon is usually one-third the thickness of the door. (The panel is very often one-third the thickness of the door too, but this is variable, although usually at least one side of the panel is recessed from the frame.)

The lock rail, being of greater depth, should have two tenons at each end, as shown below.

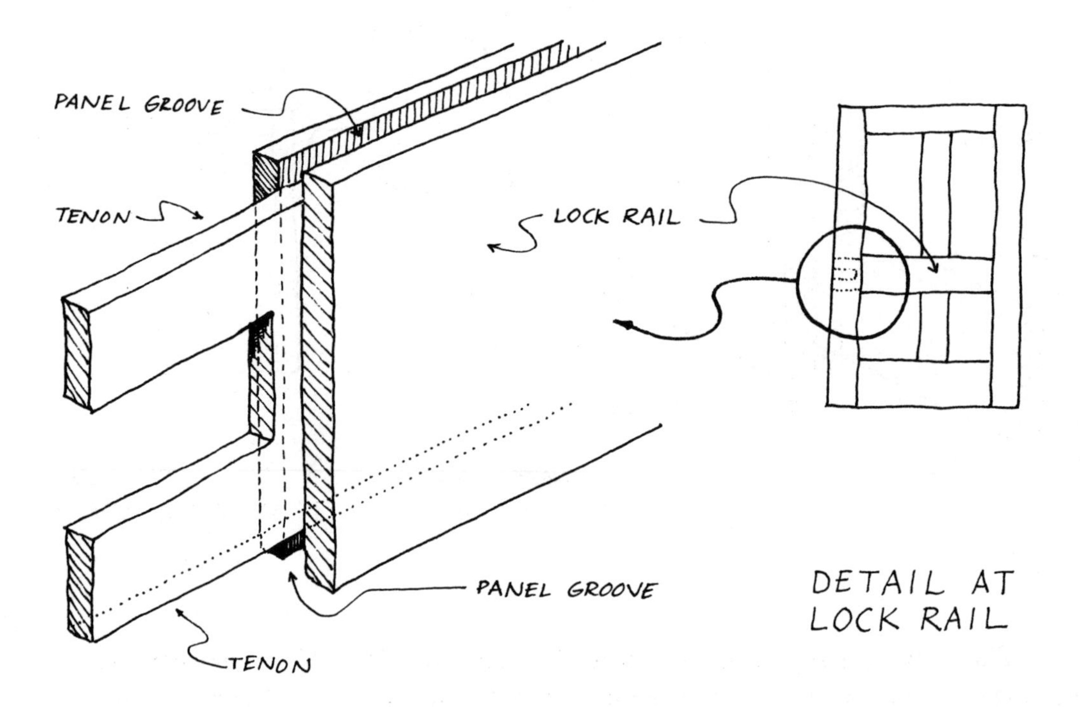

Remember that the bottoms of the two panel grooves in the rail are the top and bottom limits of the tenons.

The bottom rail should also have two tenons at each end, but the bottom of the lower tenon should be at least 1½" (**38 mm**) from the finished bottom of the door to prevent damage in wedging-up (as with the top rails). (See illustration opposite.)

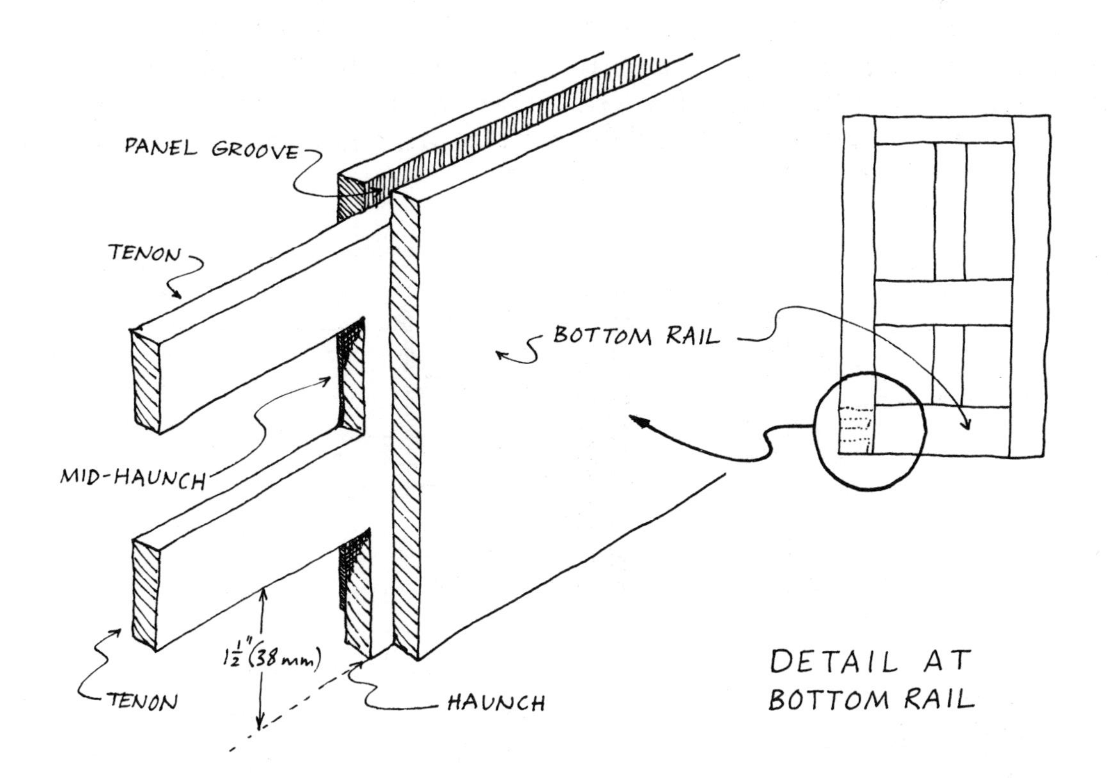

*The tenon at each end of the muntins may be as wide as the muntins themselves, less the depth of the panel grooves, of course, and about **2" (50 mm)** long, as shown below.*

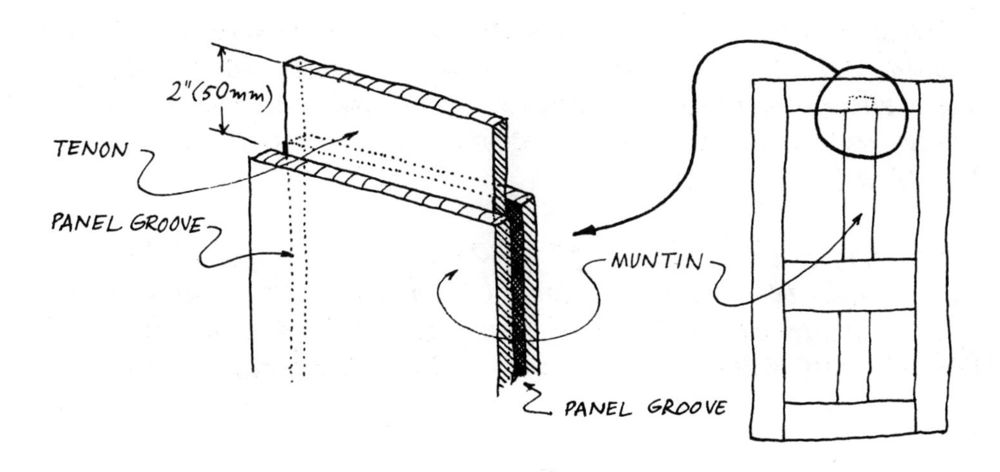

The mortises in the stiles must have the same width as the corresponding tenons in the rails, and the depth on the **inner** side must also equal the depth of the tenons. The top and bottom of the mortise must, however, slope upwards and downwards respectively to the outer side of the stile to allow for the insertion of the wedges.

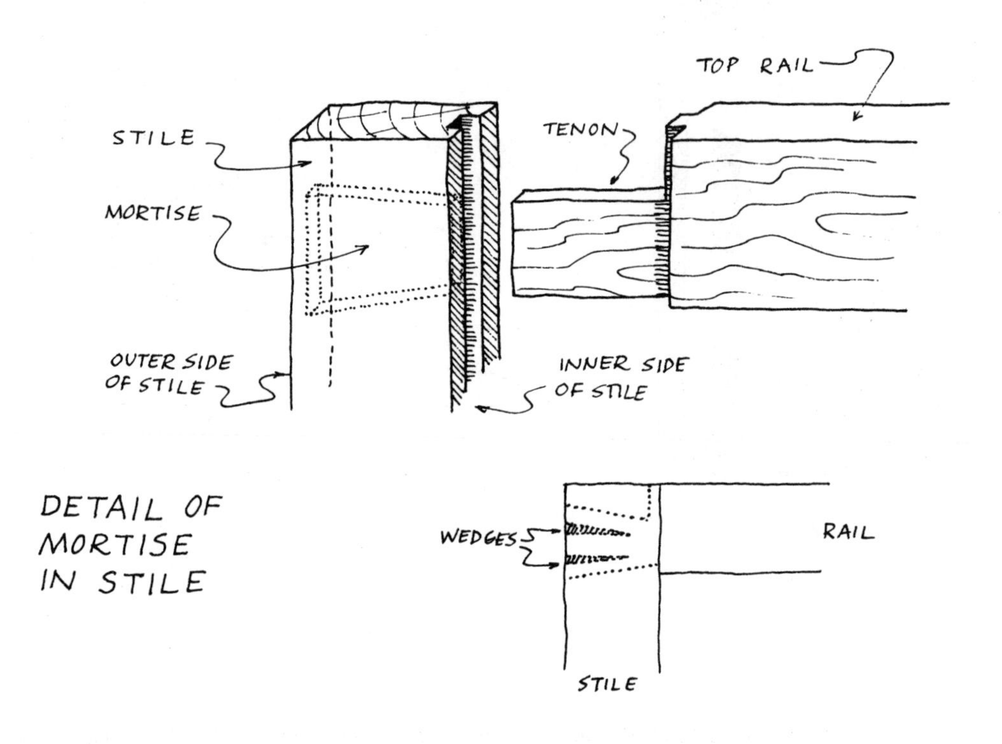

"Wedging-up" is the last process in the framing of the door and should not be done until the panels have been fitted into position, which means, of course, that **all** the joints must have first been cut and the frame "knocked up" dry to test the fit and trueness of all the joints.

For doors thicker than **2"** (**50mm**), the tenons at the ends of the rails should be double. In good construction the double tenon ought always to be made at the end of the lock rail if a mortise lock is installed at this level. The tenon at this point then becomes a double pair as shown below.

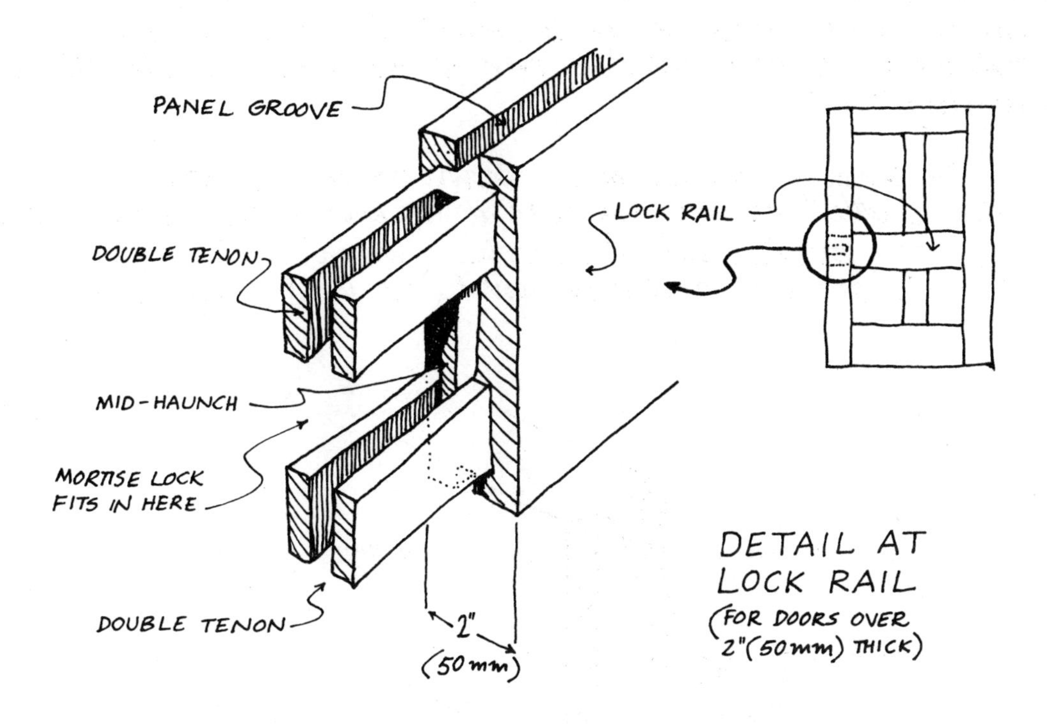

DETAIL AT LOCK RAIL (FOR DOORS OVER 2"(50mm) THICK)

For hardwood doors, and other cases where it is not desired to expose the end grain of the tenons on the edges of the stiles, fox-tail wedging is used.

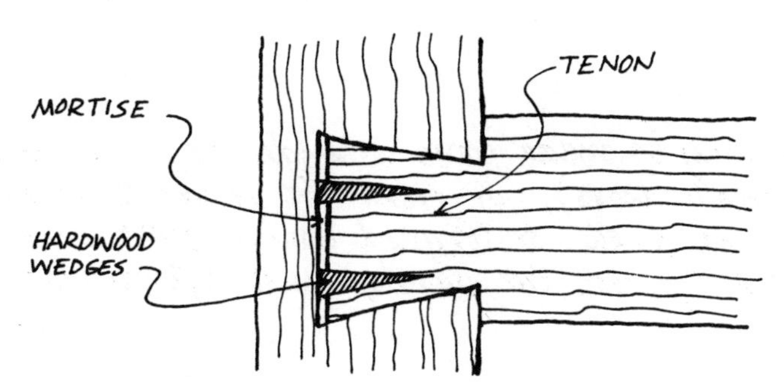

FOX-TAIL WEDGING

Fox-tail wedging proceeds as follows: The mortises are cut into the stiles to within 1" (25 mm) of the opposite edge, and the tenons cut even a little shorter to permit the shoulders of the rail to butt tightly against the stile even if the wedges should fold over in driving.

The tenon for the top rail is shown below. The wedges are inserted a little way into saw cuts at the end of the tenon and the tenon is then inserted into the mortise (which must be cut to a dovetail shape) and driven in until the joint is tight.

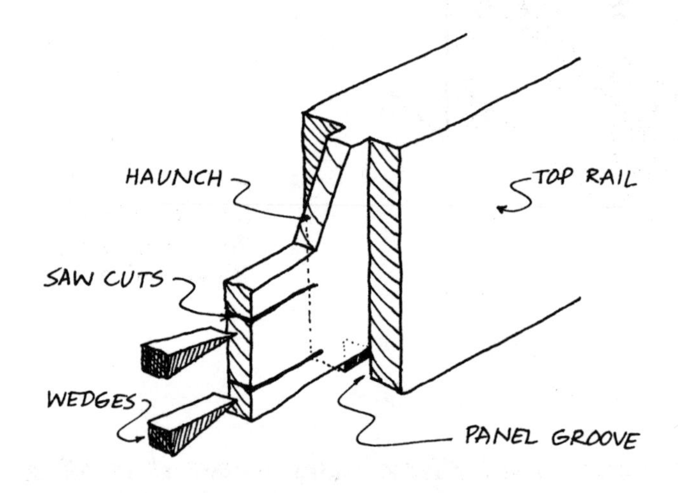

FOX-TAIL WEDGING FOR TOP RAIL TENON

Each panel may be made from a single board (plywood is often used in mill-made doors), or from two or more pieces either plain butt-jointed and glued, or tongued and grooved and glued. If made from several pieces, the panels must afterwards be planed to a true surface.

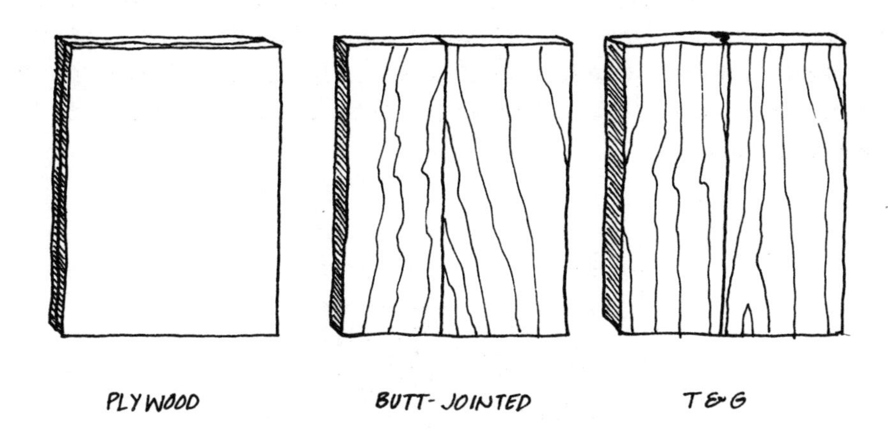

The most important point about the panels is that they must not be rigidly fixed in the framing, neither nailed nor tightly wedged, but must be free to move in the grooves so that in the event of shrinkage, they will not crack. This danger is less likely, of course, if the panels are made of plywood. Similarly, if mouldings are planted around the panels, they must be nailed to the framing, not to the panels, or the panels will either crack or pull the moulding away from the frame.

The panels may be quite plain or worked in a variety of ways. Sometimes the paneling is raised and sometimes recessed. Eighteenth-century paneling was almost always raised, the center of the panel being flush with the face of the stile, with the chamfer cut out of the solid wood and having a shallow square shoulder around the center (sometimes called the reserve). Any moulding was usually stuck rather than planted, as was the custom in the Middle Ages (and is again now).

STILE PANEL

PLAIN PANEL

RAISED PANEL IN STUCK MOULDING (18TH CENT.)

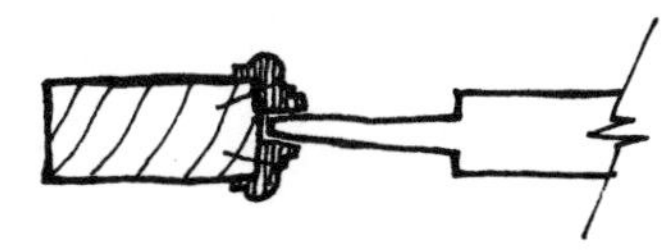

RAISED PANEL IN PLANTED MOULDING

Fine hardwood doors designed for interior use are often veneered over solid stock. This is because hardwoods are very expensive and have a greater tendency to twist and warp. Veneering is more suitable for the paneling and may be applied in a patterned manner.

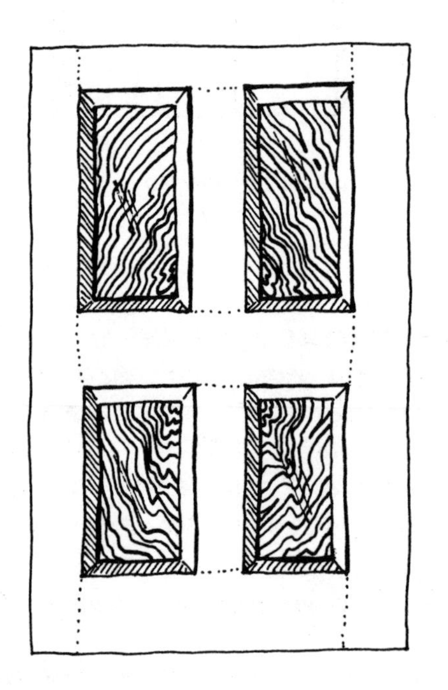

VENEERED PANELING

INSTALLATION

Although exterior doors, being usually larger and heavier, are set in frames more substantial and somewhat more complicated than frames for interior doors, both types have the same basic parts: a frame, consisting of lintel, jambs with stuck or planted stops, and a threshold; and architraves, usually referred to as casings, which trim the frame into the wall.

Since the lintel is technically a horizontal member over an opening (such as a doorway, window, or arch) designed to support the superincumbent weight, the piece at the top of a door frame which connects the two jambs is now usually called the head jamb, since, although "jamb" implies a vertical member (it comes from the French word for "leg"), it supports nothing but its own weight – the weight above the opening being supported by a "header" in the wooden framing.

Older construction invariably rabbeted out the jambs to provide a stop for the door, but modern construction usually makes a plain frame and then fixes a separate stop for the door – as will be explained further on.

The main difference between interior and exterior doors is the sill, which all exterior doors must have. In general, hinged doors should open in the direction of natural entry, especially in the case of exterior doors, which are often with a screen or storm door (which, of course, must open outwards, contrary to the general rule).

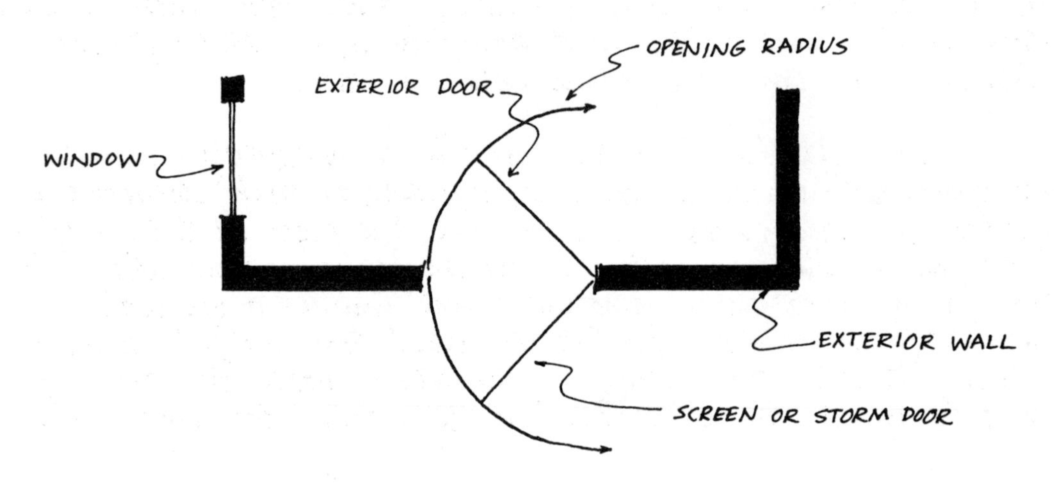

This inward opening is more difficult to weatherproof than an outward opening and has occasioned a variety of sills designed to shed water outwards and prevent its entering the house. It must therefore slope downwards and outwards. The threshold may be an integral part of the sill, as shown at A, or attached separately, as shown at B.

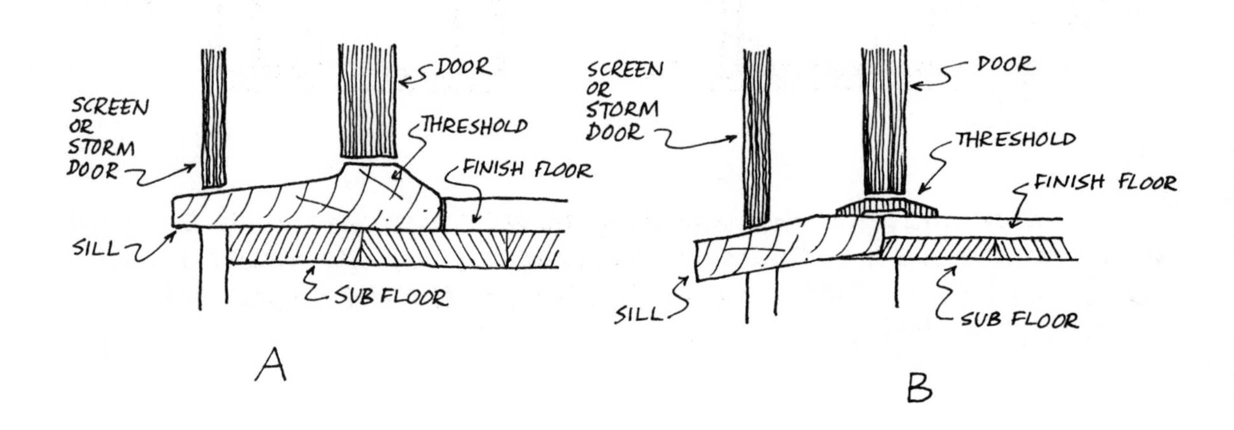

The sill (and threshold) should be of hardwood to better resist wear, and indeed most mill-made sills which may be bought at lumberyards are made of oak.

It is usual to place the sill between the rough opening, over the sub floor, and bring the jambs down to the sill.

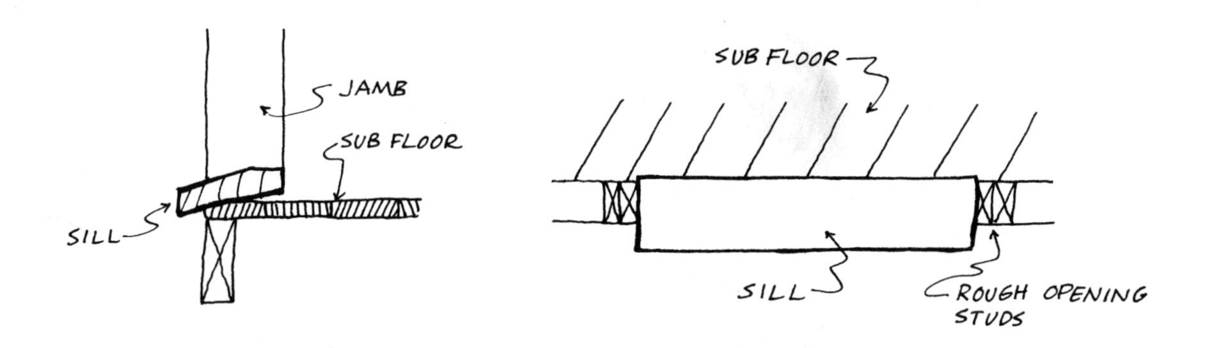

If the threshold is an integral part of the sill, it must be the width of the door thickness and beveled slightly at the front to allow the front of the door to project a little, forming a drip. Also, the back of the sill should have a lip to project over the finish floor, hiding the joint.

SILL WITH INTEGRAL THRESHOLD

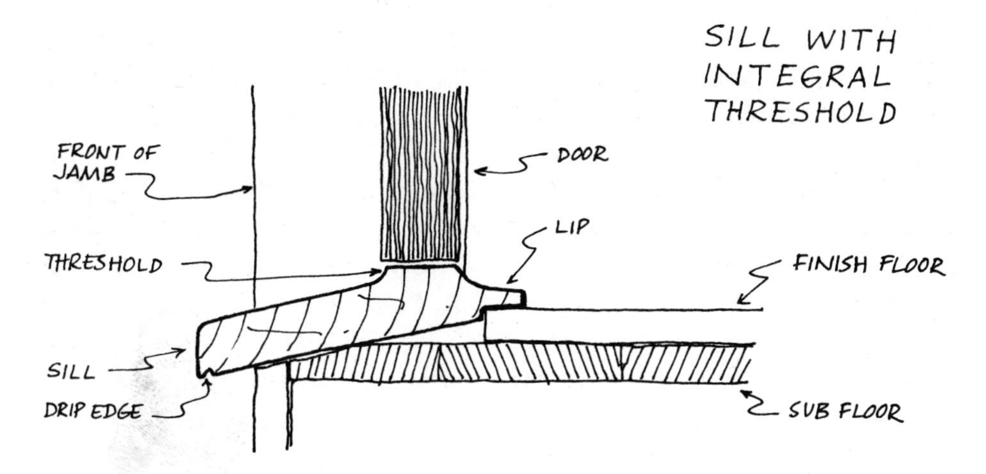

However, very often the threshold (or saddle, as it is sometimes called, because it straddles the joint between the two finish floors, if they be laid separately, of adjacent rooms) is fixed separately after the jambs have been installed. In this case it is cut to the width of the finished opening and should be undercoated or caulked to form a watertight seal between sill and finish floor.

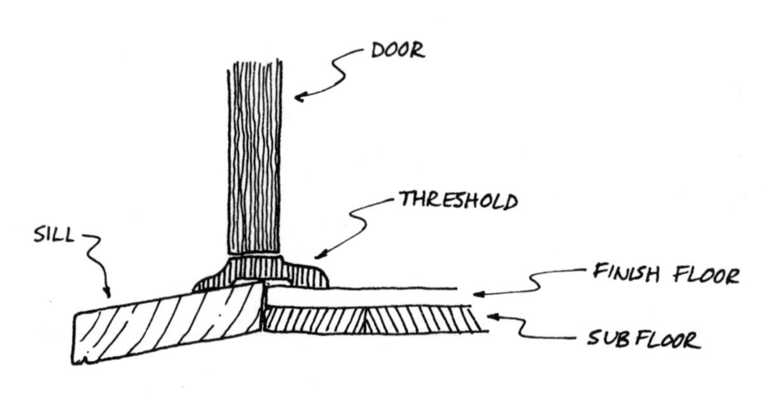

SILL WITH SEPARATE THRESHOLD

Like exterior frames, interior frames may also be set on the sub floor and have the finish floor fitted around them. This method secures the frame firmly, but it is also possible to fix the frame to the finish floor. Details of sills, thresholds, or saddles may vary from house to house, but, provided you accommodate the basic requirements of slope and sealing, all will be well.

Once you have determined what will happen at the base of the opening, you may set the frame. This should be wide enough so that its outside edges are flush with the finish surfaces of both sides of the wall in which the frame is located. This enables the trim to cover the joint as shown at A, opposite. However, sometimes the outside siding is butted up against the trim, rather than running under it. In this event the frame must be flush with the outside of the sheathing, as shown at B, opposite.

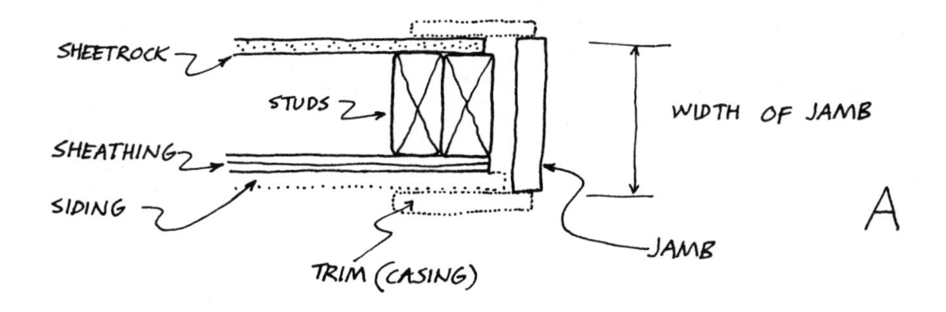

JAMB WIDTHS FOR EXTERIOR DOORS

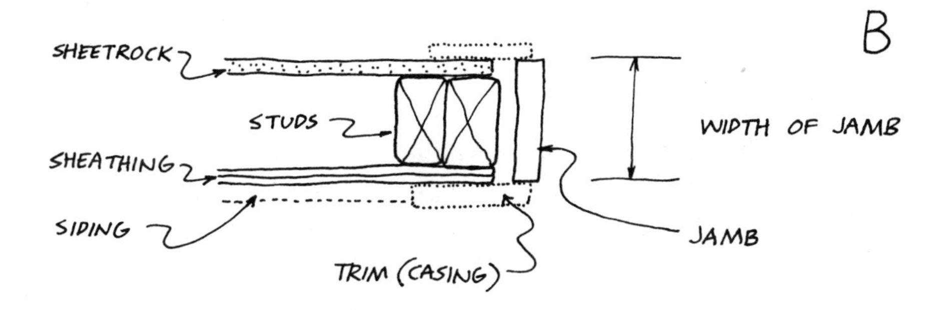

NOTE: For interior doors the jamb is made as at B, since there is not siding (the sheathing being another wall of sheetrock).

The rough opening should have been made large enough to accommodate the door, its frame, and an inch or so of blocking between the frame and the rough opening.

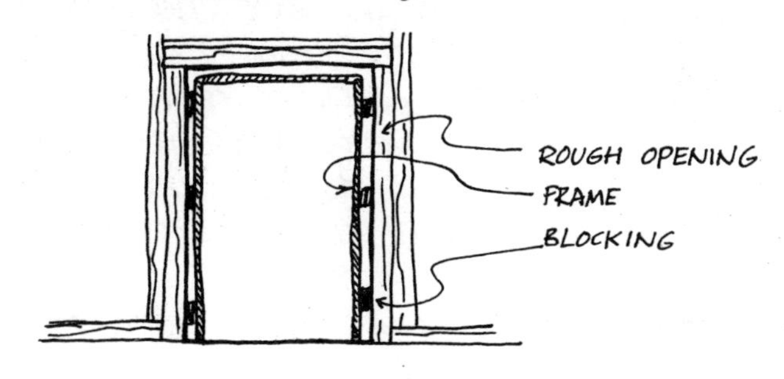

The main thing about installing the frame is that, regardless of the rough opening, the frame must be installed perfectly plumb and square. Therefore, make your frame (with the side jambs rabbeted* into the head jamb as shown below) in such a way that the head jamb will fit nicely between the studs of the rough opening, and so that there will be enough room behind the jambs to adjust for plumbness.

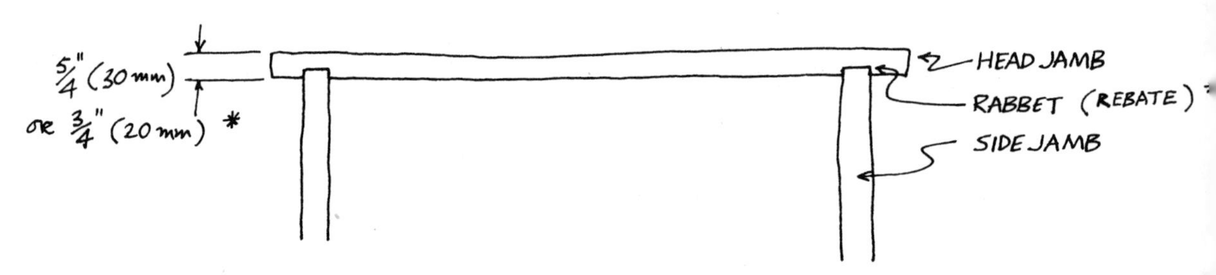

* The **5/4" (30 mm)** stock should be used for exterior doors and good quality interior doors, but **3/4" (20 mm)** stock is often used in cheaper work.

NOTE : If one of the studs of the rough opening is <u>perfectly</u> plumb, then one of the side jambs may be fixed at the <u>end</u> of the head jamb. In this way the side jamb may be fixed directly to the rough opening and made more secure. But it must still be squared forwards and backwards. If <u>both</u> studs are <u>perfectly</u> plumb, then fix that jamb which will carry the hinges.

If the floor is perfectly level, the jambs may be of equal height, but if the floor slopes from one side of the opening to the other, one jamb will necessarily be longer than the other (in order that the head jamb may remain perfectly level). If the floor slopes backwards or forwards through the opening, the bottoms of the jambs will have to be cut accordingly.

* See explanatory note on page **66** at end of chapter.

Whatever the case may be, make sure that the head jamb is level (it is easier to fix it if it fits snugly in the rough opening), and the side jambs reach the floor – although you do have a little leeway if you are installing the frame onto the sub floor, since the finish floor will cover some of the gap.

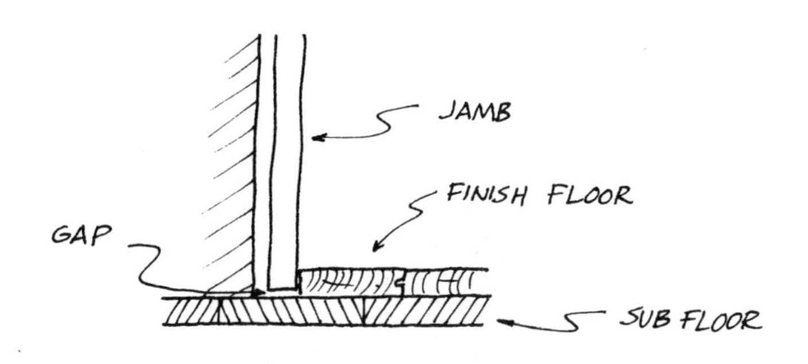

When you are satisfied that the head jamb is level, the side jambs must be made plumb and fixed to the rough opening. One easy way of doing this is by using wedges made from shimming shingles, a bundle of which is very useful for the house under construction. After the jambs are plumb – and square – that is, they protrude from the rough opening the correct amount – on <u>both</u> sides of the opening – nail through the jambs and the wedge blocking into the studs.

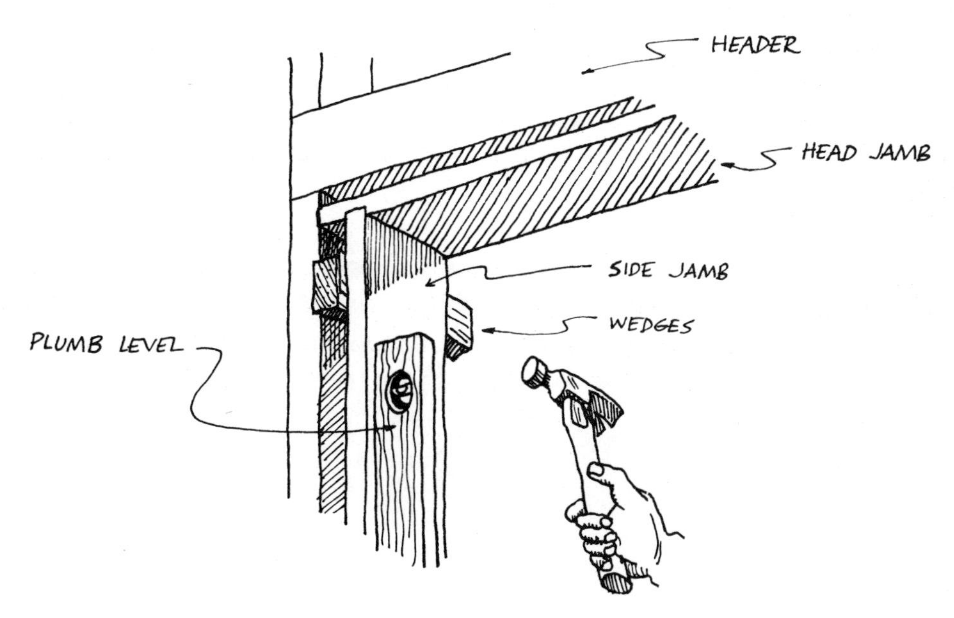

If the door stop is to be applied separately, then nail the jamb at a point where the stop will cover the nail head! Also, it helps to toenail the bottom of the side jambs from both sides, as shown below.

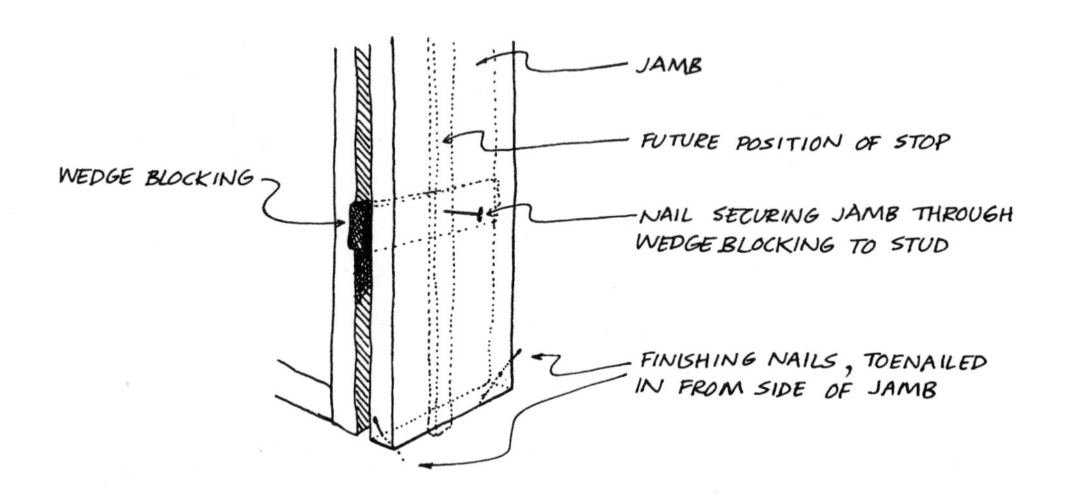

While an integral stop is better security, since it prevents the insertion of a blade behind the stop to push back certain kinds of lock bolts, a separate stop facilitates the hanging of the door and also makes it possible to change the door's swing from one room to another if this should ever be desired. If you ever want to do this with an integrally rabbeted stop, you will have to cut the door down all around and install a separate stop on the frame.

In any event, it is usually best to hang the door at this stage and fix the stop (if it is separate) to the frame with the door in position.

If the door is new, the stiles will have to be sawn off flush. Then, having measured the width of the finished opening, plane the hinge stile perfectly smooth and square to an amount that will equal the lock stile, so that the two stiles are equal in width and there is $\frac{1}{16}$" (2mm) clearance either side of the door.

The lock stile should be planed at a slight angle back to the stop to enable the inside edge of the door to clear the jamb when opened.

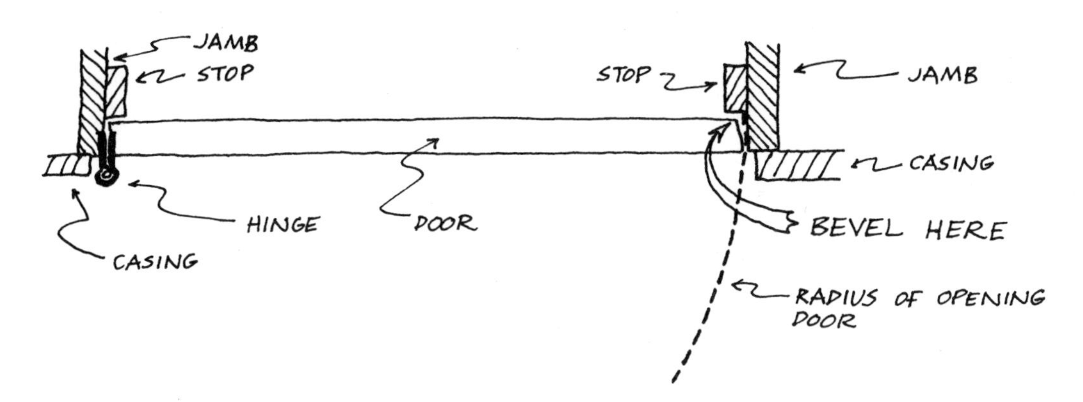

There should also be $\frac{1}{16}$" (2mm) clearance at the top of the door, and $\frac{1}{8}$" (4mm) clearance at the bottom if there is a threshold, and maybe more if the door will have to clear a rug or carpet.

If the door is very oversize, it is best to leave as much of the bottom rail as possible, shortening the door from the top, since, if you take away too much of the bottom rail, the door will eventually fall apart. (This applies only to frame doors.)

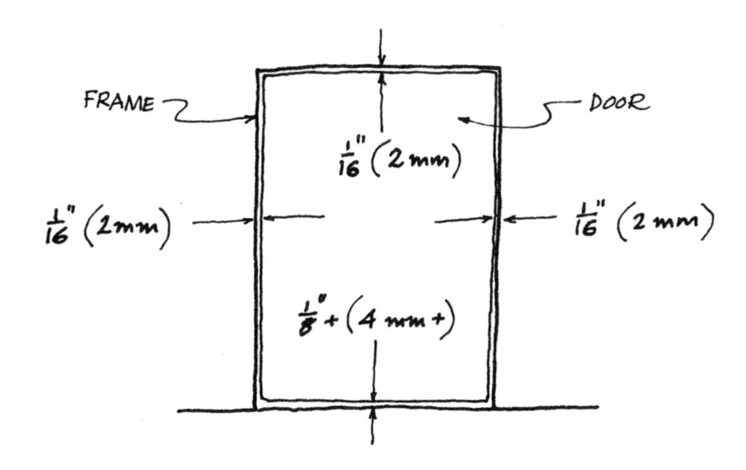

Wedge the door in the opening so that there is the correct clearance at top and bottom, but make the hinge stile butt tightly up against the hinge jamb. Three hinges are best, the lower being about 10" (250 mm) from the floor, the upper about 6" (150 mm) from the head jamb, and the middle one centered between the other two. However, if you are hanging a ledged and braced door, the hinges should be fixed to the ledges.

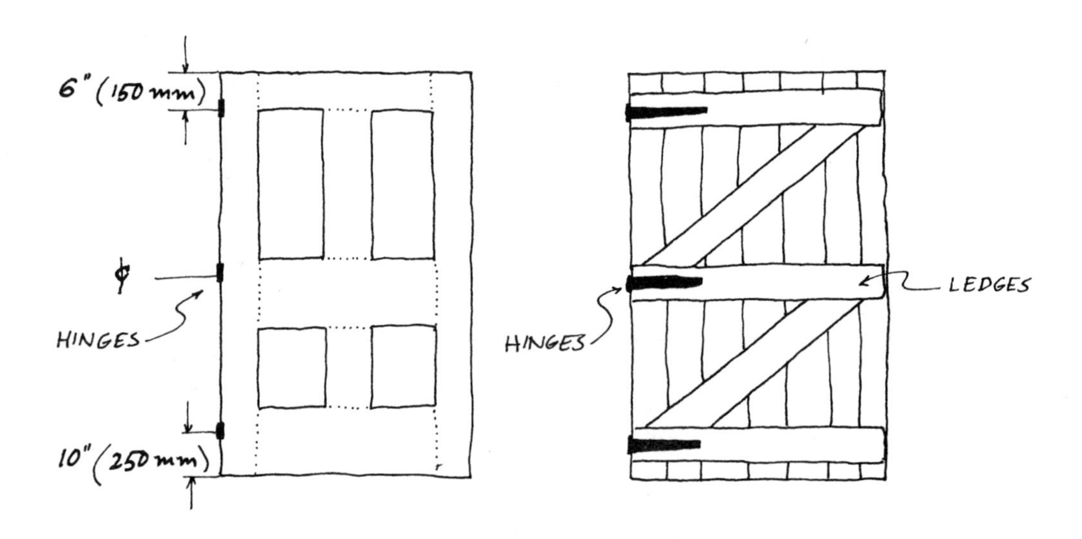

Hinges are as old as doors, and consequently there is an amazing variety. Most frame doors are today hung on loose pin butt hinges which are mortised into the edge of the door and the face of the jamb.

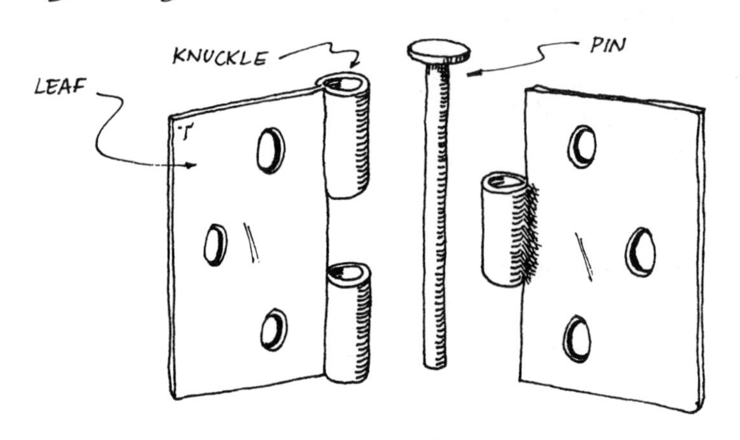

LOOSE PIN
BUTT HINGE

Most ledged and braced doors are hung on "tee" or "T" hinges, which are made in a variety of patterns and known by a number of names, some of which are illustrated below. These hinges are face mounted.

CROSS-GARNET OR T-HINGES

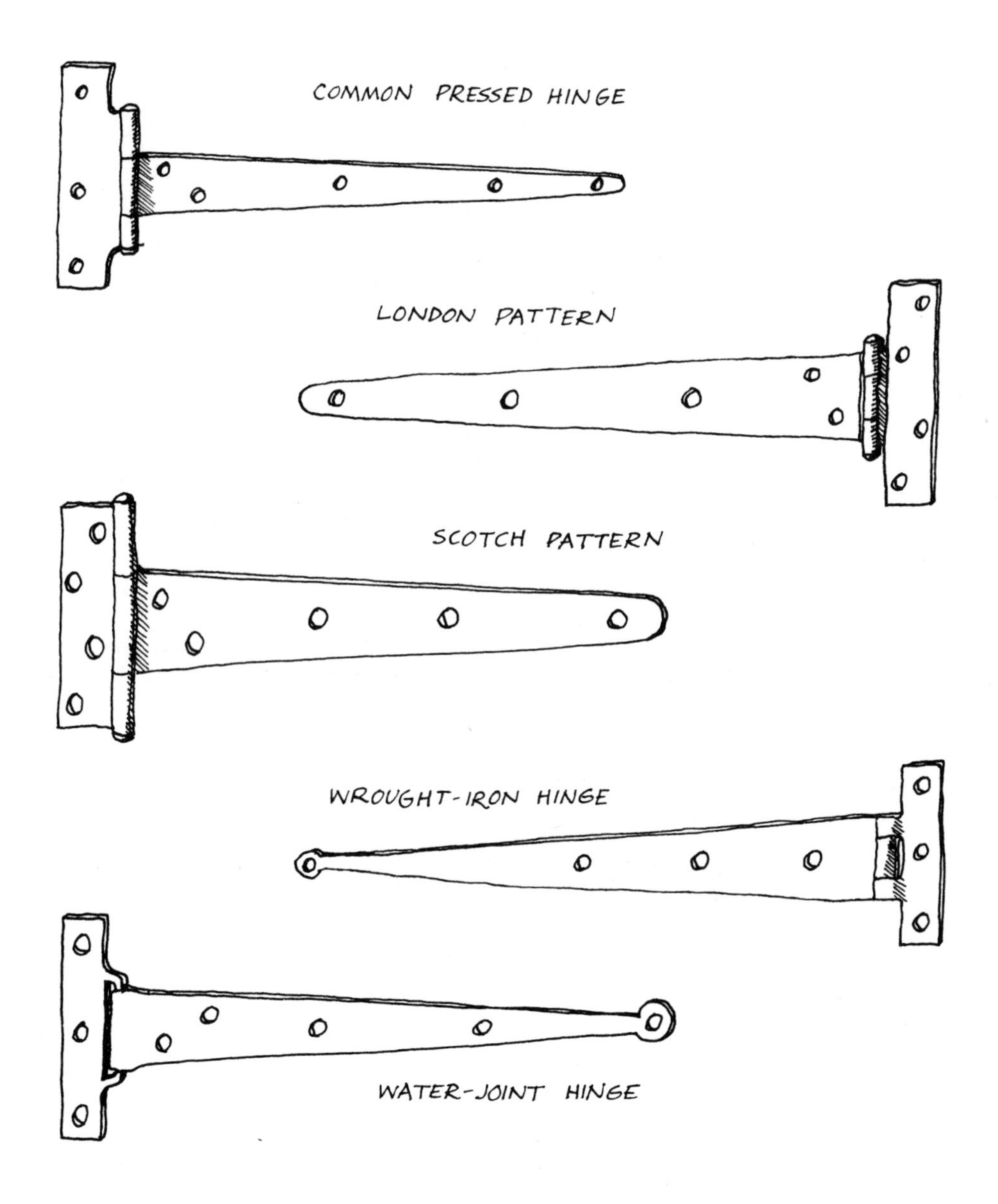

There are also a number of special-purpose hinges such as table-leaf hinges and invisible cabinet hinges too specialized to be discussed here. But one variety of specialty hinge deserves mention here—the rising butt hinge. This raises the door a slight amount when it is opened so that the door will close again by its own weight.

RISING BUTT HINGE

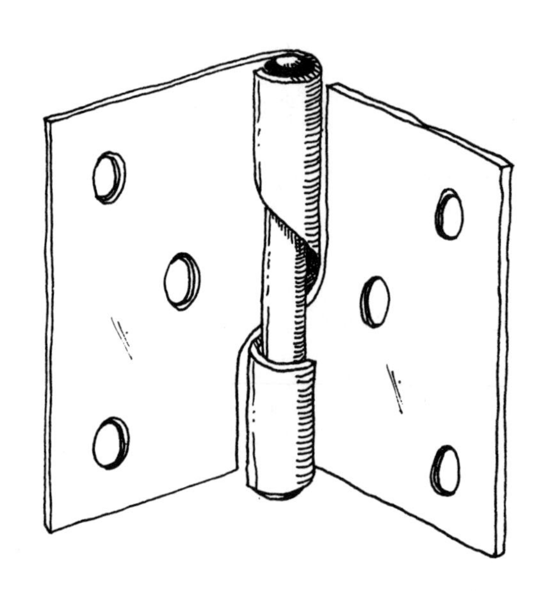

A rising butt hinge will necessitate a slight bevel at the top of the door to allow it to clear the frame as it rises during opening.

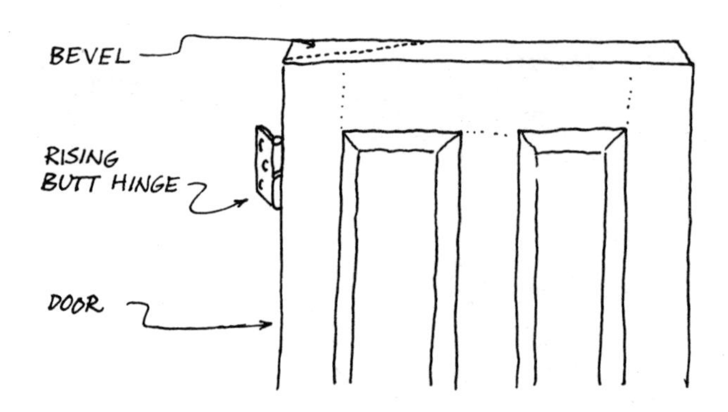

Another advantage of the rising butt hinge is that it allows the door to clear thick carpets and yet not have a large gap at the bottom of the door when closed.

Older hinges were often made flat, whereas modern hinges are usually swaged – the difference is shown below.

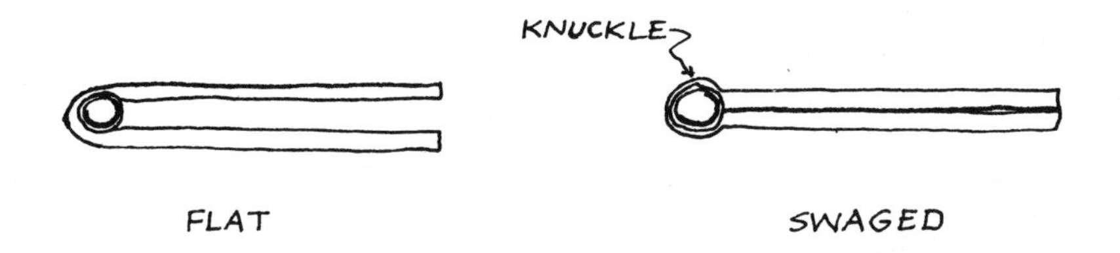

Swaged hinges may be mortised equally, whereas flat hinges must be offset.

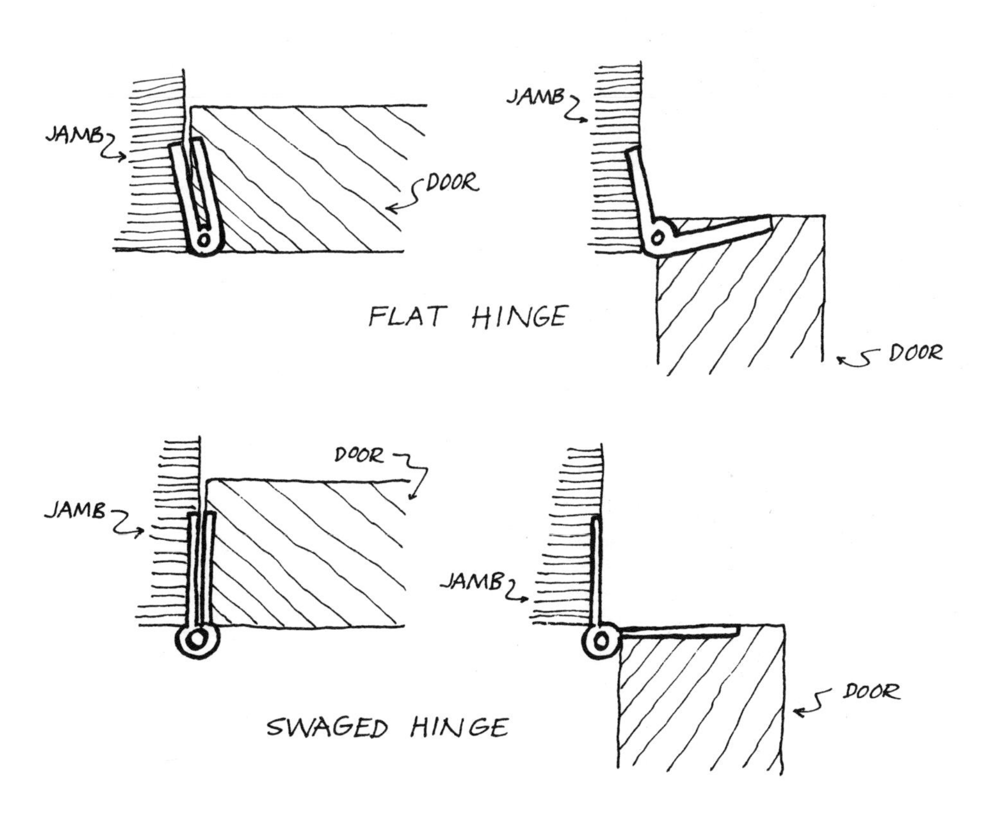

Assuming you are hanging a frame door with swaged butt hinges, the commonest operation, the procedure is as follows:

Mark the position of the hinge leaves on the door and the jamb, bearing in mind that if you want the door to open flat against the wall, the hinges must be set out a little so that the door will clear the casing.

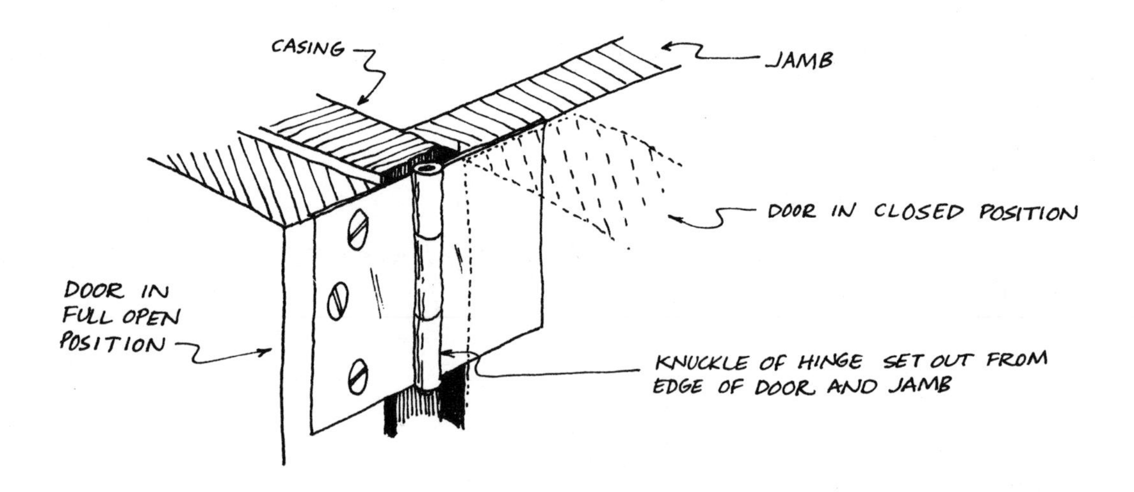

The leaves of the hinges must fit flush into rabbets cut especially for them. The hinge is stronger if it has three bearing surfaces (as shown below) rather than extend right across the door.

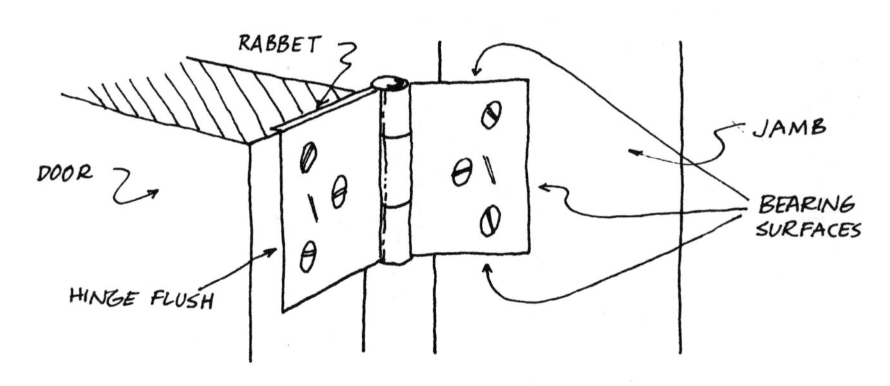

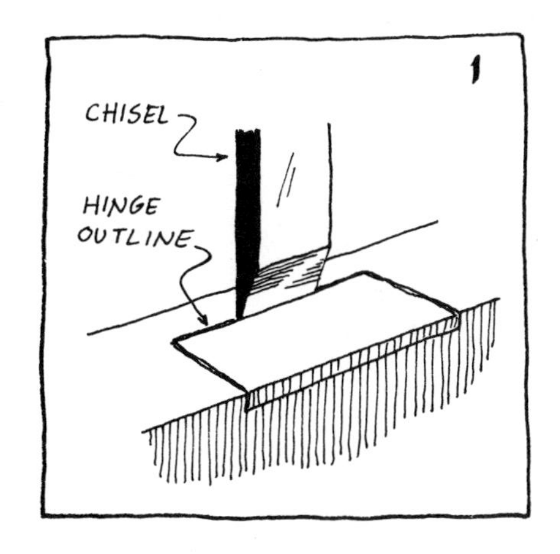

Separate the leaves of the hinges and fix each leaf with one screw only into its respective rabbet. This makes adjustment of the leaves easier when meshing the hinge knuckles.

When the door is hung, check that it opens and closes freely. If it doesn't shut all the way, you have set the hinges too deeply – in which case insert a thin piece of card behind the leaf. When you are satisfied, insert the remaining screws, but leave the pins sticking out a little in case the door must be removed again.

The door stop may now be fixed to the jambs, with the door wedged shut.

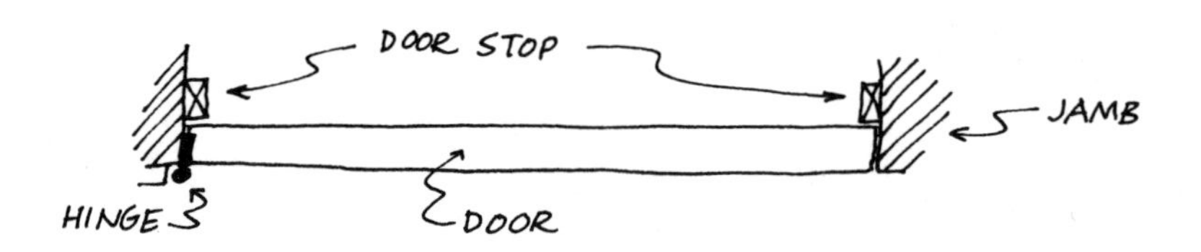

You may use a mill-made stop, which is usually moulded on one side and which should therefore be mitered at the top, or you may make up your own stop from 1" × 2" (20mm × 50 mm), in which case it is better to butt the ends.

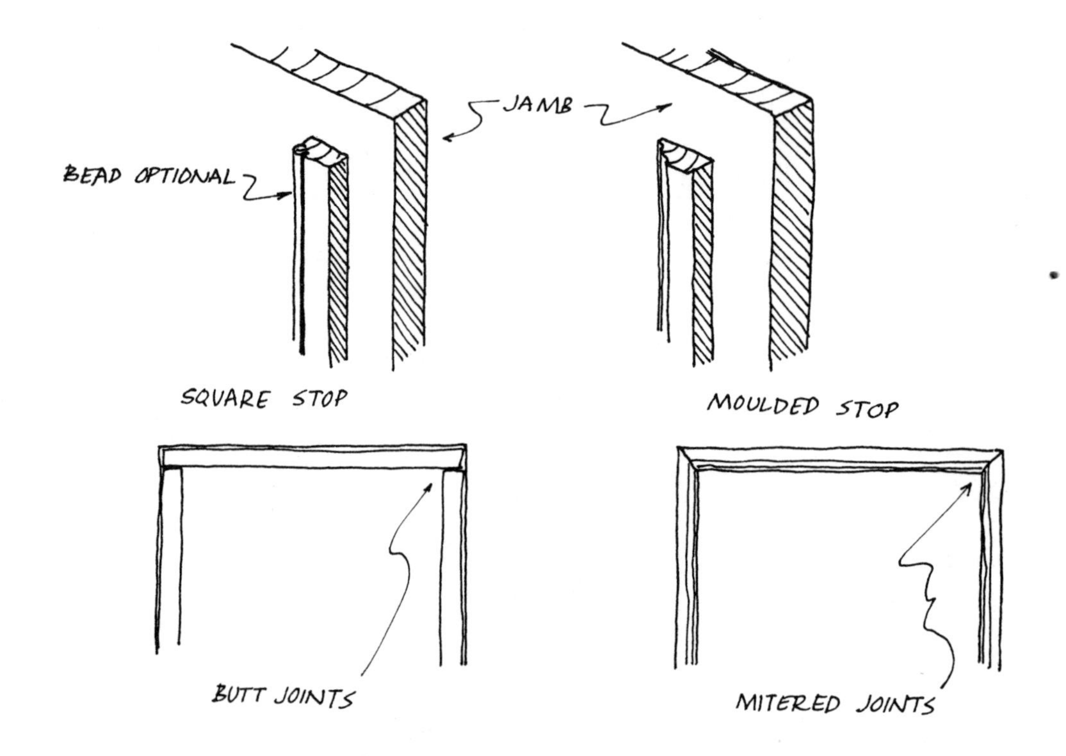

The door trim or casing may similarly be bought (in a variety of patterns) or made up on the job, and may be anything from a very plain board to an ornately moulded architrave complete with a cabinet head.

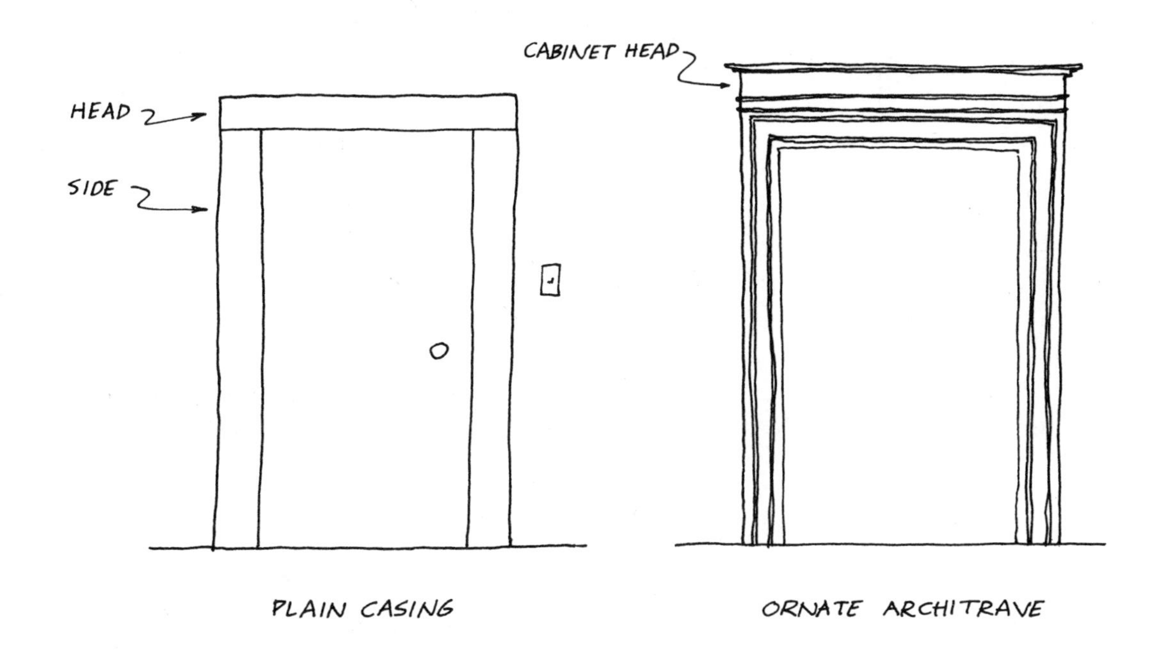

PLAIN CASING ORNATE ARCHITRAVE

If you use any mill-made casing with a moulded edge, such as clamshell, you will have to miter the joint between the head casing and the side casing, but this is unfortunate, since, even if this joint is made with a spline, it almost inevitably shrinks and leaves an unsightly gap.

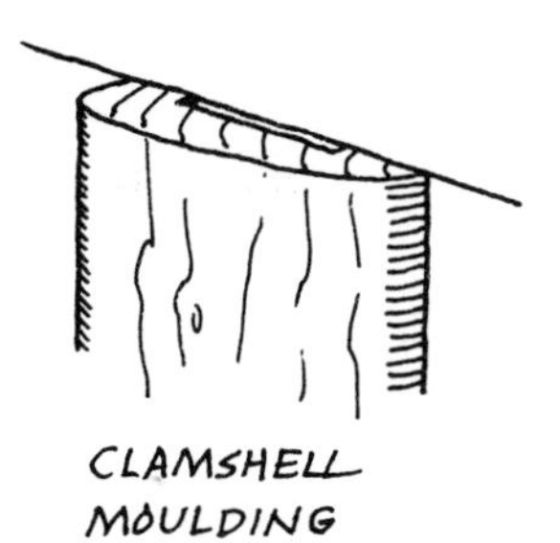

CLAMSHELL MOULDING

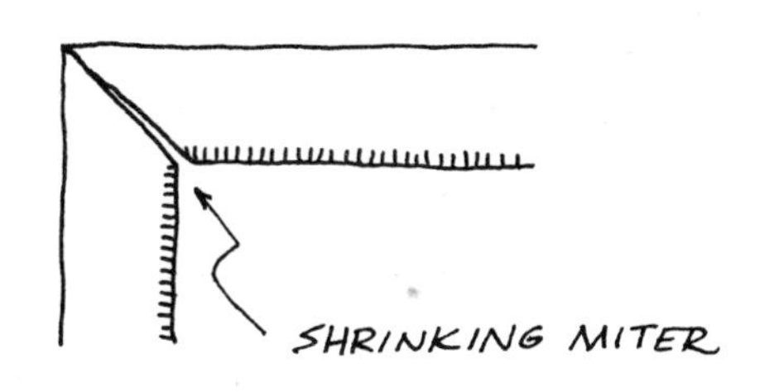

SHRINKING MITER

It is therefore better to use a plain casing and make a butt joint, adding any moulding you desire separately.

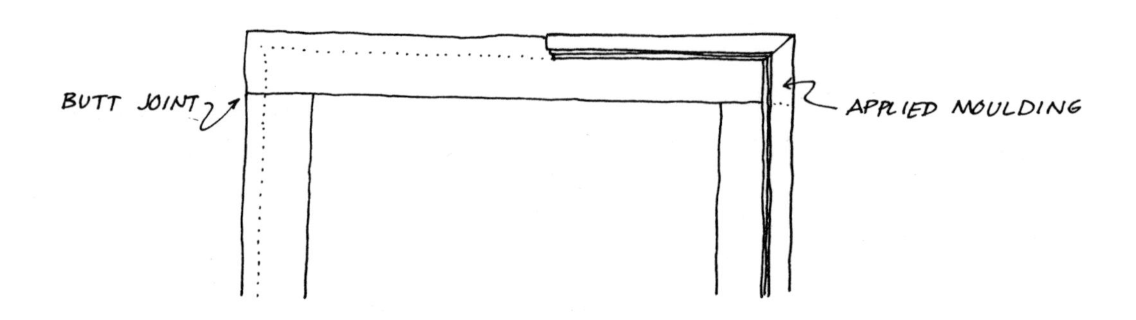

In this case, fix the side pieces first, setting them back an equal amount all around from the edge of the jambs.

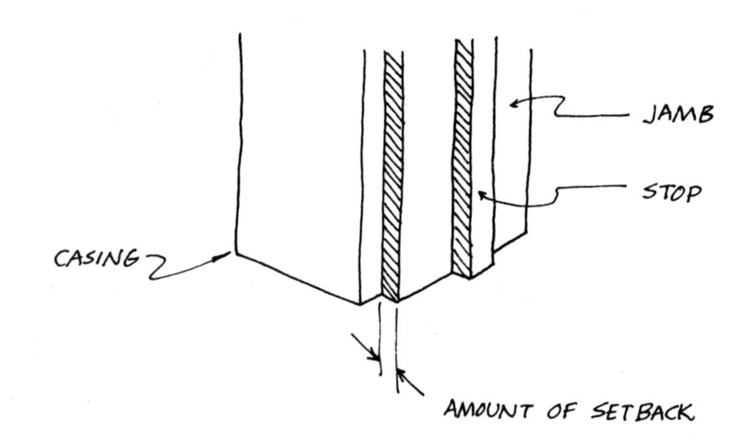

The head piece may then be measured and fitted exactly.

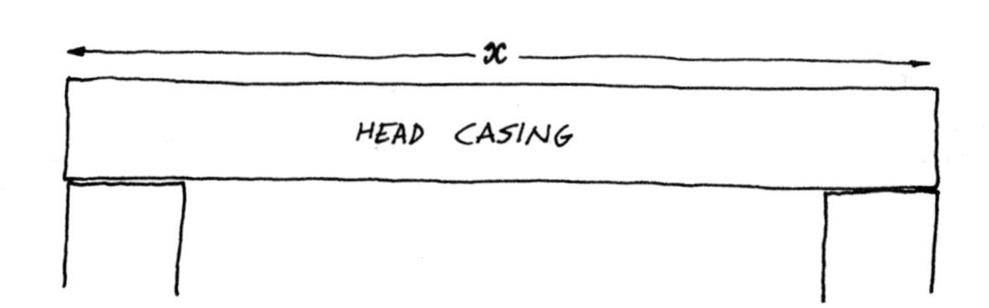

Be careful when nailing the jamb edge of the casing not to pull the jamb out of line. Also make sure that you nail the other side of the casing into a stud.

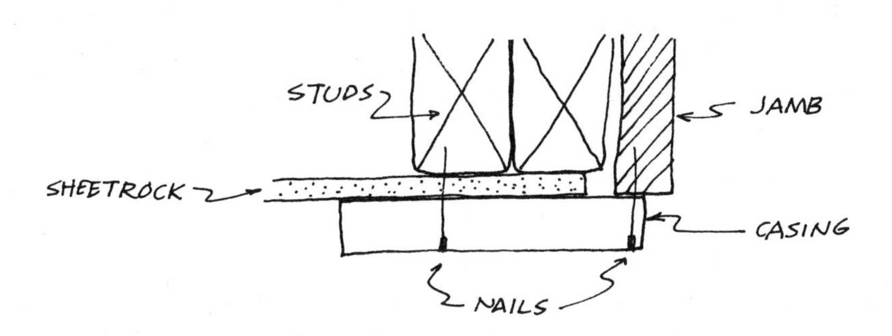

Furthermore, when spacing the nails, be sure not to nail where the lock or striker plate will be located.

Most of the remaining hardware, such as locks, handsets, knockers, spy holes, letter slots, etc., comes in package form, usually with installation instructions and templates, and it is beyond the scope of this book to describe every type.

RABBET VERSUS REBATE

Considerable confusion exists on both sides of the Atlantic regarding the words "rabbet" and "rebate." The British talk about "rabbet" planes and "rebating," the Americans talk about "rabbeting," and "rabbets" themselves are spelled "rebate" in Britain, though commonly pronounced "rabbet."

All these words came into English from the French word "rabattre," meaning to beat down or reduce. Since the spelling "rebate" is commonly used in Britain and America to refer to a financial repayment, I have thought it useful to limit the woodworking applications of the word to the older spelling "rabbet."

Consequently, I have used "rabbet" for noun, adjective, and verb throughout this book; e.g.:

a rabbet, instead of a rebate (British)
a rabbet plane, instead of a rebate plane (rare)
to rabbet, instead of to rebate (American and British variety)

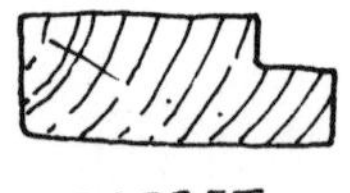

RABBET

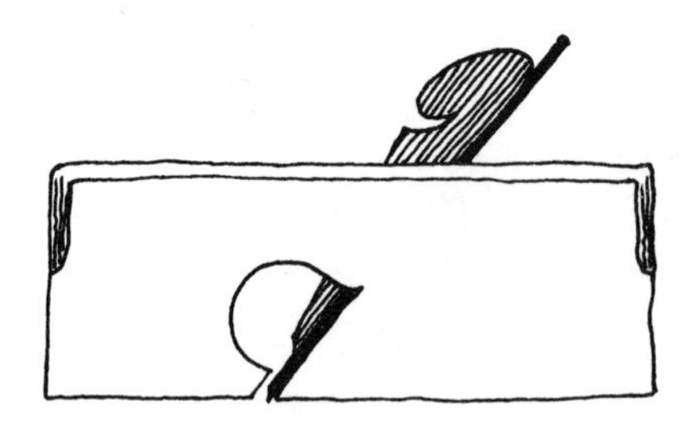

BEECH RABBET PLANE

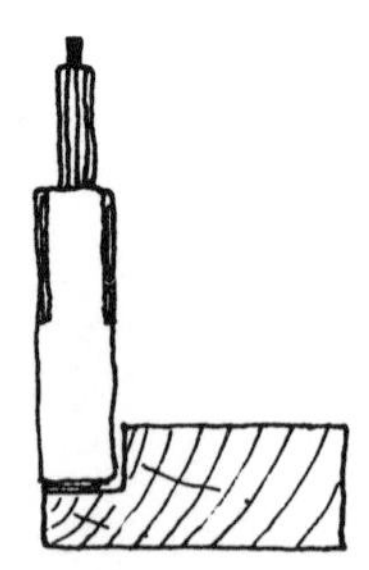

RABBETING

Chapter Two

Windows

All windows consist of two basic parts: the glazed frames, called sashes (and sometimes casements), and the frames or cases of various kinds which contain these. Windows are generally classified according to the manner in which the sash is held in the frame; these classifications may be described as follows:

1. FIXED WINDOWS
2. CASEMENT WINDOWS
3. DOUBLE HUNG WINDOWS

Before describing the construction of the various types in detail, it should be pointed out that all kinds of windows may be obtained as ready-made units from a number of manufacturers in many different shapes, styles, and sizes. They may be made of wood or metal, and most come with frame, sash, and trim assembled. Often they are glazed and have the appropriate screens, weatherstripping, and hardware in place. Furthermore, they may be made with double glass, which is much more efficient as a thermal barrier than single panes, both for keeping the house warm in winter and cool in summer.

They are, of course, much more expensive, unit for unit, than other windows, but most professional builders now prefer them because they save so much time – and so, for the contractor, so much money. All that need be done is to order the right size in the appropriate style, place it in the rough opening, and when the inside walls are finished and the outside walls are sheathed, position the unit and secure it in place by nailing the trim or casing to the window frame and the inside and outside walls.

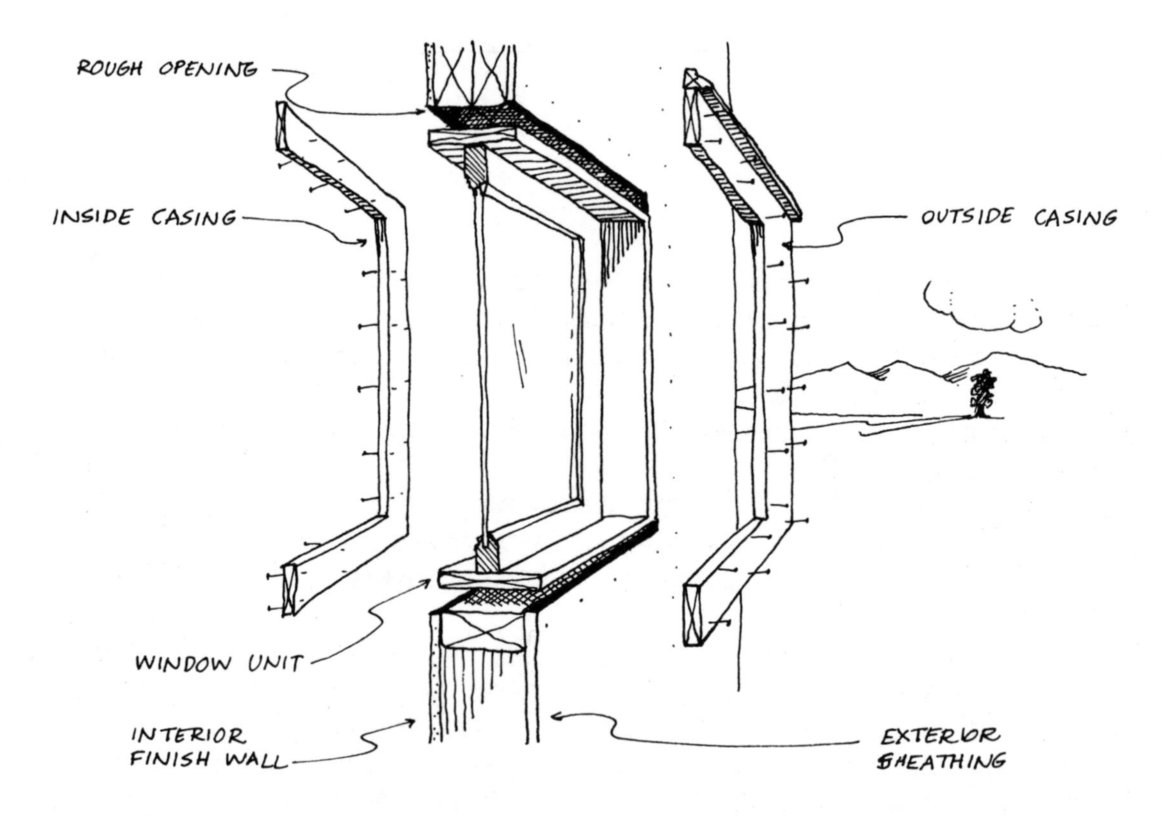

PREFABRICATED WINDOW INSTALLATION

1. FIXED WINDOWS

Theoretically, fixed windows are not really windows at all, since the word "window" comes from a Saxon word meaning "wind eye" and referred to a hole in the house, often the roof, through which not only light, but also air, wind, and, most important in those days, smoke could pass and fresh air enter.

The only requirement for successful construction of fixed windows is that the glazing be properly done and the sash be secure and watertight in its frame, which may usually be effected by good flashing and caulking.

FIXED WINDOW

2. CASEMENT WINDOWS

Casement windows have sashes which may be hung to swing inwards or outwards, hinged from the top (called awning sash), from the bottom (called hopper sash), or from the sides. The window may consist of one sash or a double sash, or even two sashes and a fixed central sash (the last arrangement being applicable to double hung windows also).

Casement windows have the advantage of being able to be opened completely - and of catching a parallel breeze and slanting it into the room.

If they open inwards, as is usual in Europe, the fixing of screens and storm sashes is facilitated, and both sides of the glass may be cleaned from within the room. If they open outwards, as is more usual in Britain and America, screens and shutters must be fixed on the inside - which makes it difficult to open or close the window without special hardware - which so often gets stuck! However, it is easier to make outward-opening casement windows watertight, particularly against driving rain and snow.

CASEMENT WINDOW

3. DOUBLE HUNG WINDOWS

The double hung window consists of two sashes which slide past one another, usually vertically. These windows have some advantages and some disadvantages. The advantages are that screens and storm sash can be located on the outside without interfering with the operation of the window, and ventilators and air conditioners can be placed easily in the window. The disadvantages **are** that only one-half of the window area may be opened at any one time, and that construction is a little more involved than for casement windows.

DOUBLE HUNG WINDOW

NOTE: This type of window is often confusingly referred to as a 'sash' window – all windows have sashes! What is really meant is a window, the sash of which is balanced by sash weights, which are described further on, in contradistinction to casement windows, the sashes of which are hung on hinges.

CONSTRUCTION

Factory-made window units, as mentioned before, come complete and require only to be put in place and secured there, usually by the inside and outside trim. Such convenience is, however, costly, and for the owner-builder, for whom time is not as expensive as it is for the contractor-builder, there are distinct advantages in dealing with windows differently.

One possibility is to obtain secondhand units from breakers' yards or secondhand building supply firms. It is often possible to find really handsome, well-made windows at a fraction of their cost when new. What is more difficult is to find exactly the right size or sufficient of the same sort.

A more practical approach for someone who has time, and desires to save money, and still have a well-made window, is to buy secondhand sash and make the frames. It is possible to find sash in almost any size and pattern, and providing the stiles and rails are big and strong enough, they may be hung in almost any fashion you desire.

Again, the sash is the part which holds the glass. If there are several panes of glass in the sash, it is described as a four-light, six-light, or twelve-light window, according to however many panes there are. The frame is the part that holds the sash, and includes the casing, which holds the frame in the wall.

FOUR-LIGHT WINDOW

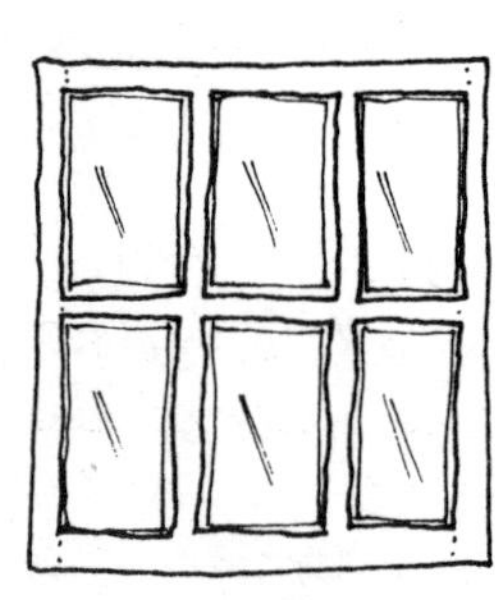

SIX-LIGHT WINDOW

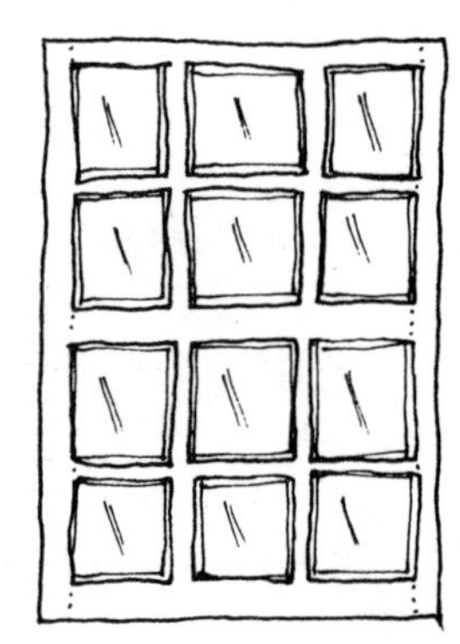

TWELVE-LIGHT WINDOW

SASH

1. CASEMENT WINDOWS

Having obtained appropriate sashes that will fit into the rough openings with **4" to 6"** (**100 mm to 150 mm**) space around all sides (smaller sash can be accommodated by making the rough opening smaller), first measure the width of the rough opening as shown below.

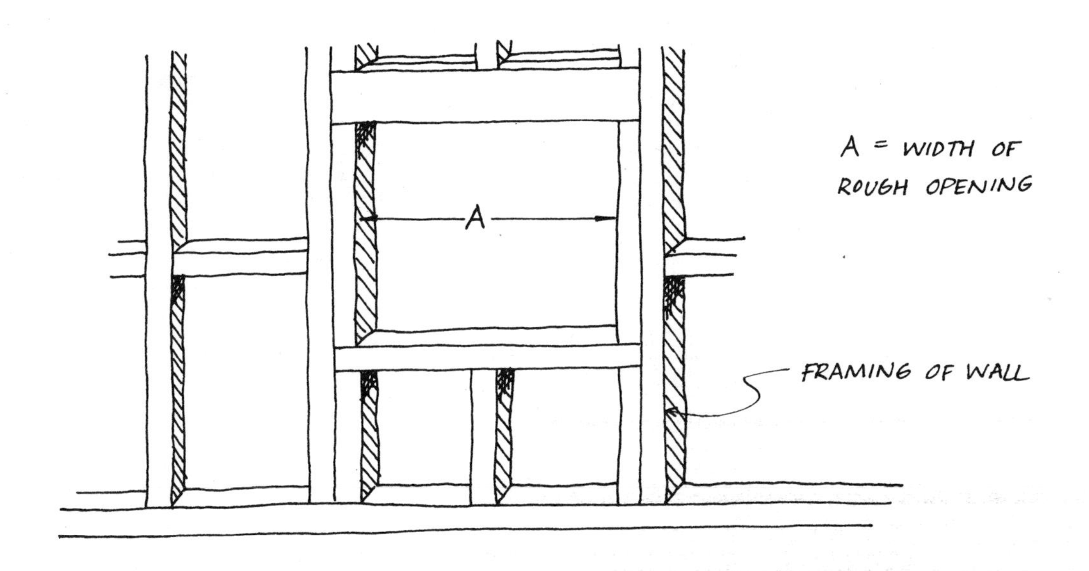

Next, cut a piece of **5/4"** (**30 mm**) lumber to be the head jamb of the window frame as shown below.

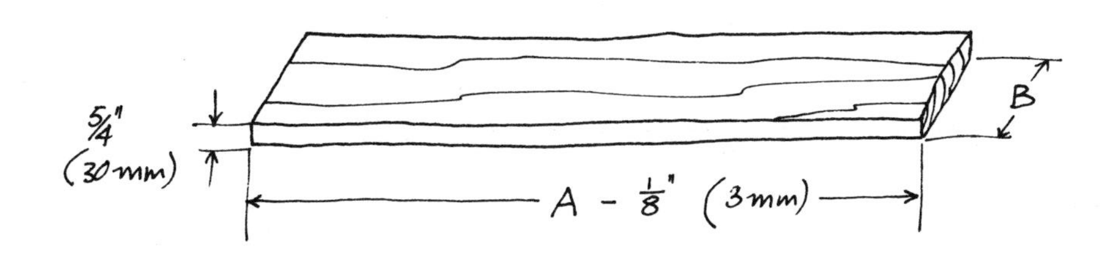

The length equals A – the width of the rough opening less

an 1/8" (3 mm) in order that the head jamb will fit easily into the rough opening.

The width, B, equals the width of the studs, plus the thickness of the exterior sheathing and the interior finish wall, plus another 1/8" (3 mm) or so for irregularities.

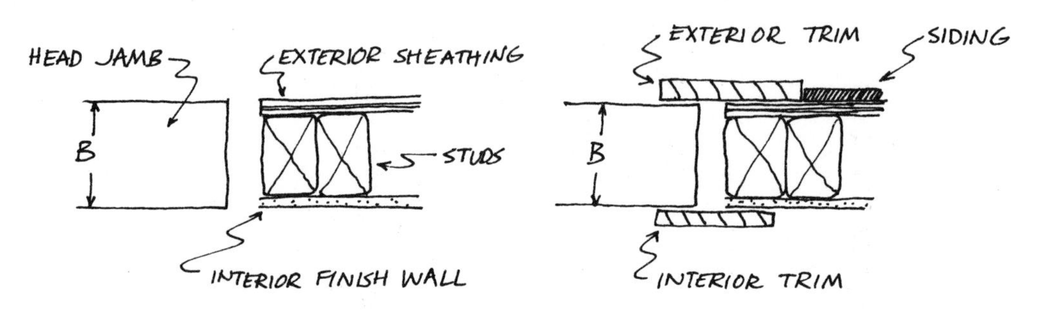

SHOWING WIDTH OF HEAD JAMB IN RELATION TO ROUGH OPENING AND INTERIOR AND EXTERIOR TRIM CASING

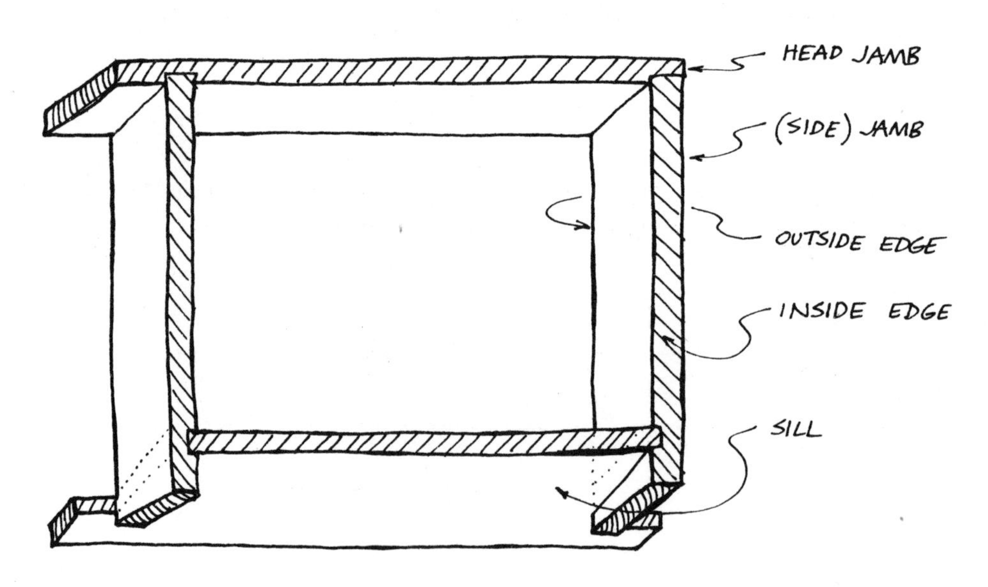

PARTS OF A WINDOW FRAME

In what will be the underside of the head jamb, cut two dadoes, each the thickness of the jambs.

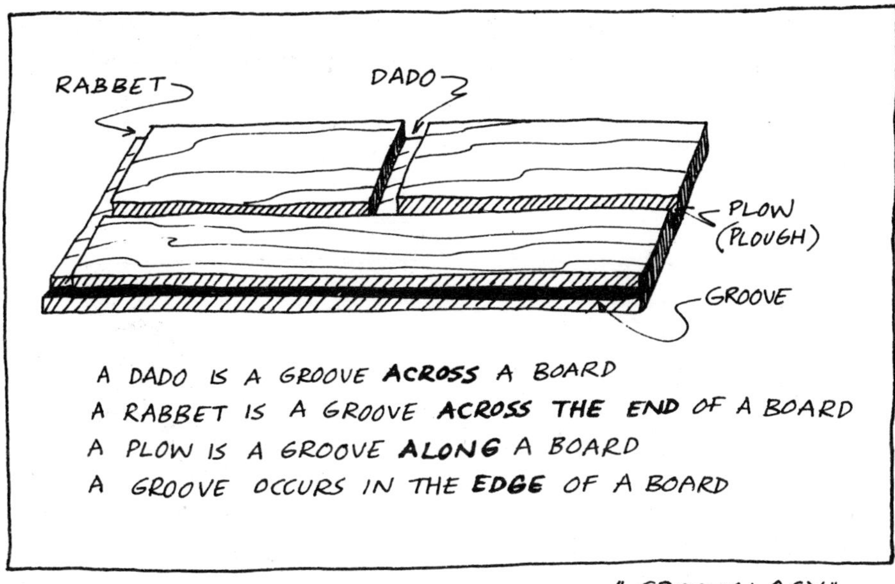

RABBET
DADO
PLOW
(PLOUGH)
GROOVE
A DADO IS A GROOVE **ACROSS** A BOARD
A RABBET IS A GROOVE **ACROSS THE END** OF A BOARD
A PLOW IS A GROOVE **ALONG** A BOARD
A GROOVE OCCURS IN THE **EDGE** OF A BOARD

"GROOVOLOGY"

The jambs, as well as the head, should be made from 5/4" (30mm) lumber; therefore the dadoes will be the same width as the thickness of the lumber being used. Measure this carefully, for you want a snug joint.

Make the first dado (actually a rabbet) at the end you want to hinge the sash. If the framing stud on this side is straight and plumb, you will make the window extra strong by fixing the hinge jamb directly to it.

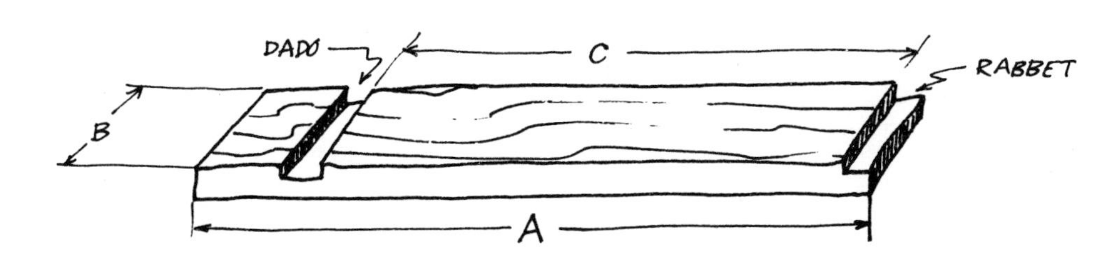

In the illustration on the previous page, C equals the width of the sash to be hung, plus a small clearance space. The dadoes should be about one-third the depth of the lumber being worked.

To make a dado, mark the edges of it carefully first, and then make several saw cuts across the board, as shown at A below, and chisel away the dado, making the bottom as smooth as possible, as shown at B below.

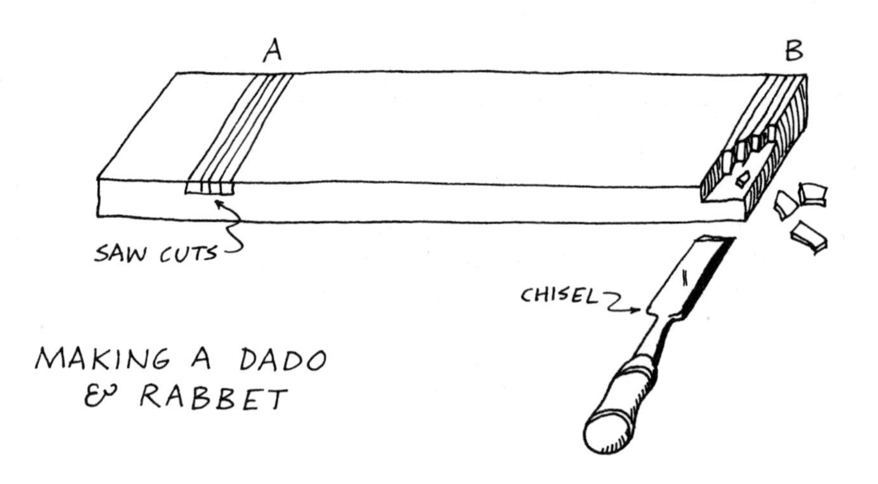

MAKING A DADO & RABBET

A circular saw is easiest if you can set the depth of the blade (or the height of a table saw blade). A "dado head set" – a special combination of blades designed for electric saws – is easiest and quickest, but not really worth the time involved in setting it up if you only have one or two dadoes to cut.

Pre-electric carpenters had planes with irons (blades) of varying widths which could be set to the required depth. However, even they usually simply used a back (tenon) saw and a chisel.

When the head of the frame is complete, the two jambs must be cut to length and dadoed to receive the sill. The length of the jambs depends on the size of the sash and the thickness of the sill. Whatever this measurement turns out to be, add a little more so that the frame you are building will stand on the rough opening at the height you desire. This, of course, is something you will have had to have taken into account when designing the window, for everything will have affected this positioning - from the dimensions of the rough opening to the size sash you are now working with.

VARIOUS ROUGH OPENINGS WITH FRAMES IN POSITION

To allow water to run off, the sill should slope downwards and outwards at an angle of about **15°**. Assuming the sash to be **hung** is going to open outwards - the commonest arrangement where English is spoken - the bottom of the sash should also be trimmed to a matching **15°**, for a close fit when shut.

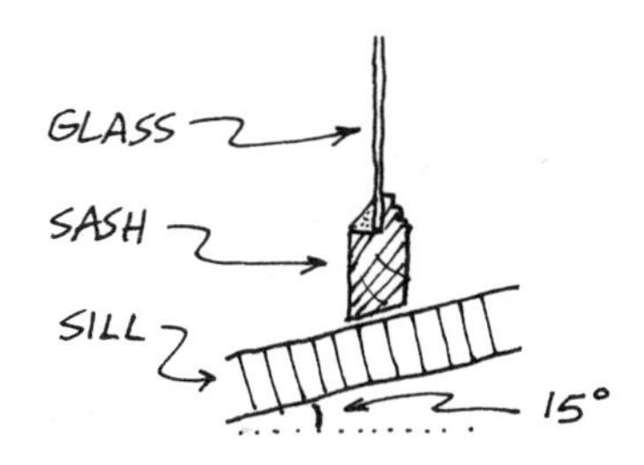

As the window is to open outwards, the sash is hinged to the outside edge of the jamb. Therefore measure in from the outside edge of the jamb the thickness of the sash, and draw a line parallel to the edge at this distance.

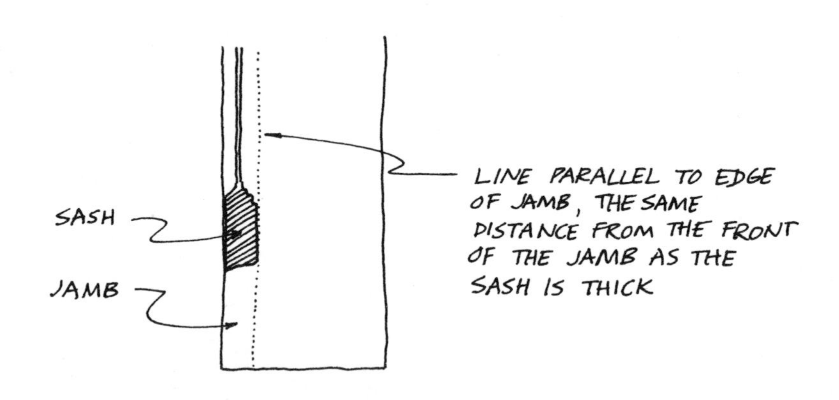

Fitting the jamb into the dado previously cut in the head, measure off the height of the sash along this line, adding ¼" (5 mm) to allow the sash to open and close freely. Draw a line at 15° across the jamb to intersect this point – and you have the top edge of the dado to be cut to receive the sill, as shown below.

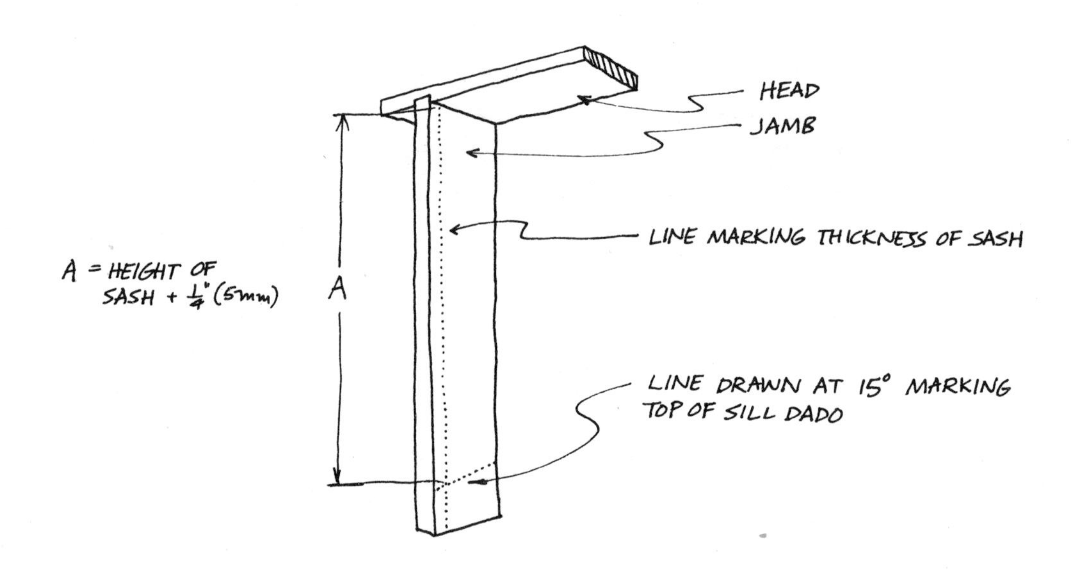

Now measure the exact thickness of the sill – it may be made from 5/4" (30 mm) stock, but better windows are usually made with 2" (50 mm) stock – and mark off this thickness below the first line and cut the dado as for the head. When this has been done on both jambs the frame should look as below.

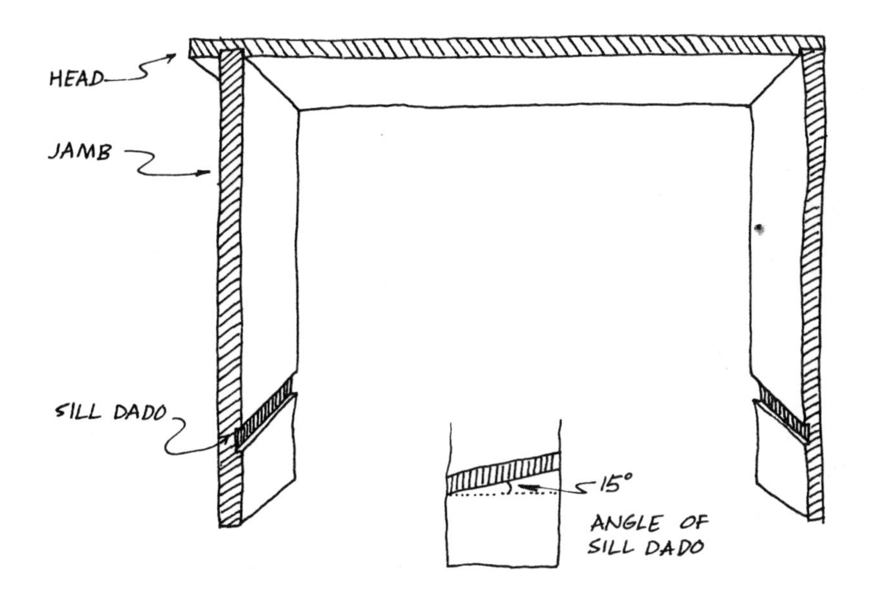

At this point you may have to trim the bottoms of the jambs so that the frame will fit easily into the rough opening – with a little to spare so that one side or the other may be shimmed up if the opening is not perfectly square.

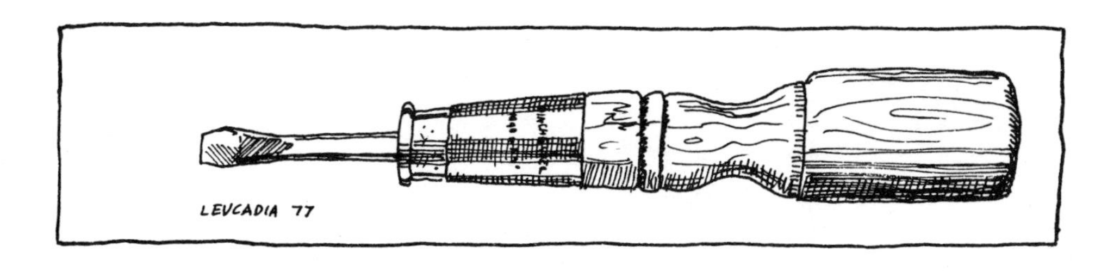

The sill should now be cut to the pattern and dimensions shown below.

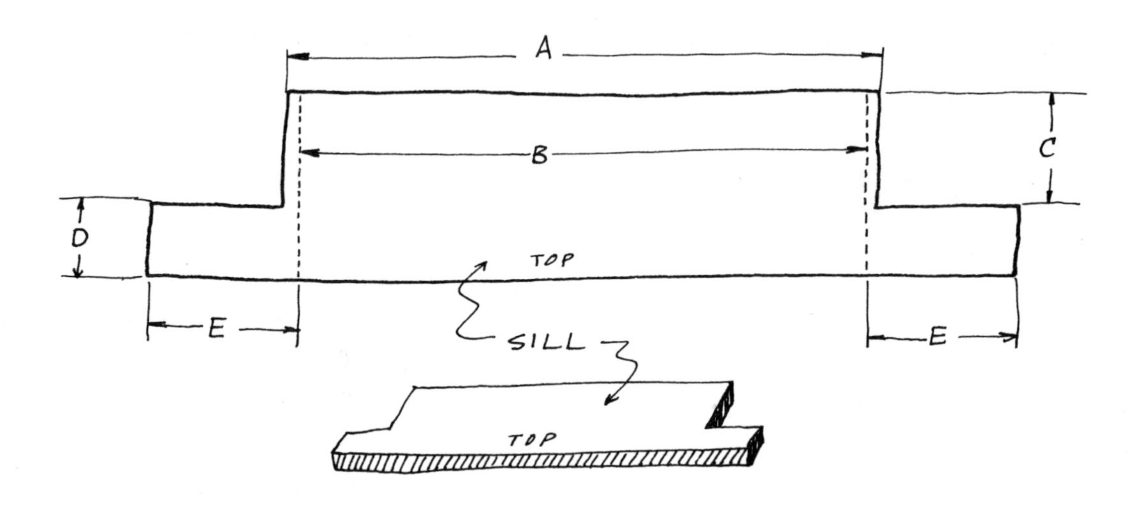

A = THE DISTANCE BETWEEN EACH DADO BOTTOM IN THE JAMBS
B = THE DISTANCE BETWEEN THE INSIDE FACES OF THE JAMBS, WHICH SHOULD EQUAL THE WIDTH OF THE SASH PLUS 1/4" (5mm)
C = THE WIDTH OF THE DADO IN THE JAMB
D = AT LEAST 2" (50 mm)
E = THE SAME AS THE WIDTH OF THE OUTSIDE TRIM (4"-6" (100 mm-150 mm

The back of the sill must be cut at an angle of **15°** in order to be vertical, and a groove cut in the under side of the front to act as a drip edge.

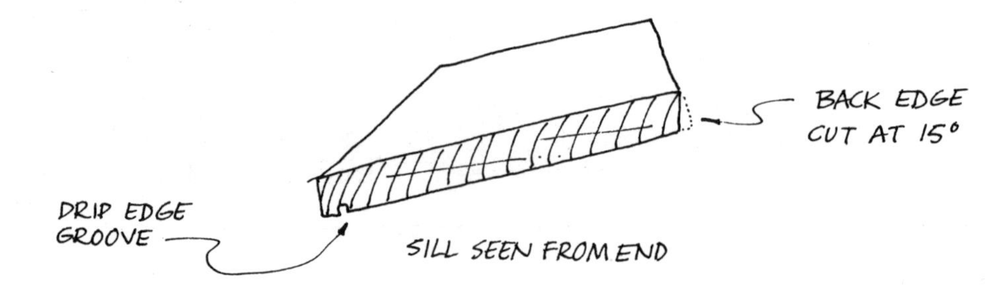

Now that all four sides have been cut, the whole thing should be nailed together and placed in the rough opening to see if it fits! When nailing take care to get the frame as square as possible, or the sash will not fit.

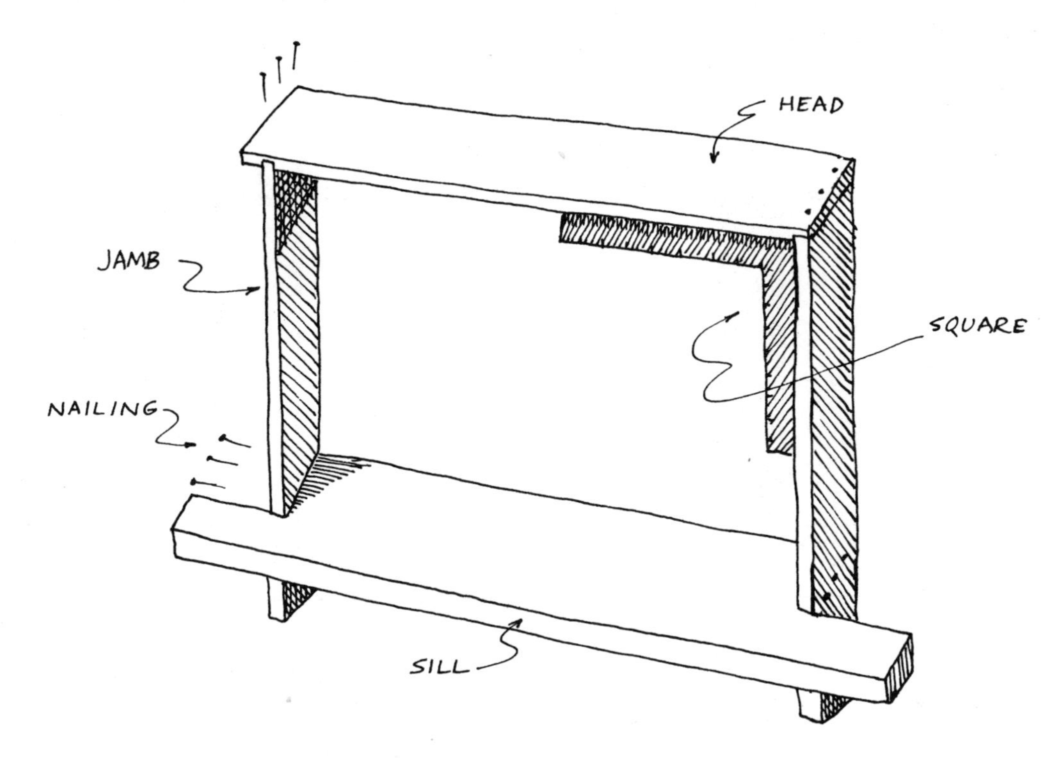

TESTING THE FINISHED FRAME
WITH A FRAMING SQUARE

The frame is held in place by nailing the outside trim to the edges of the frame and to the studs comprising the rough opening (through the sheathing, and a layer of building paper).

The trim is usually made and fixed to the frame before the whole is placed in the opening – that is, the outside trim, for at this stage in the building of a new house the inside walls are unlikely to be finished – you couldn't get the frame in the opening anyway if the trim were on both sides!

The procedure for making the trim is as follows. Cut and nail to the frame the two side pieces first, remembering to cut the bottom edges at **15°** so that the slope of the sill will be matched. The top of the trim should extend a little beyond the bottom of the head – by the same amount, in fact, that the sides are set in.

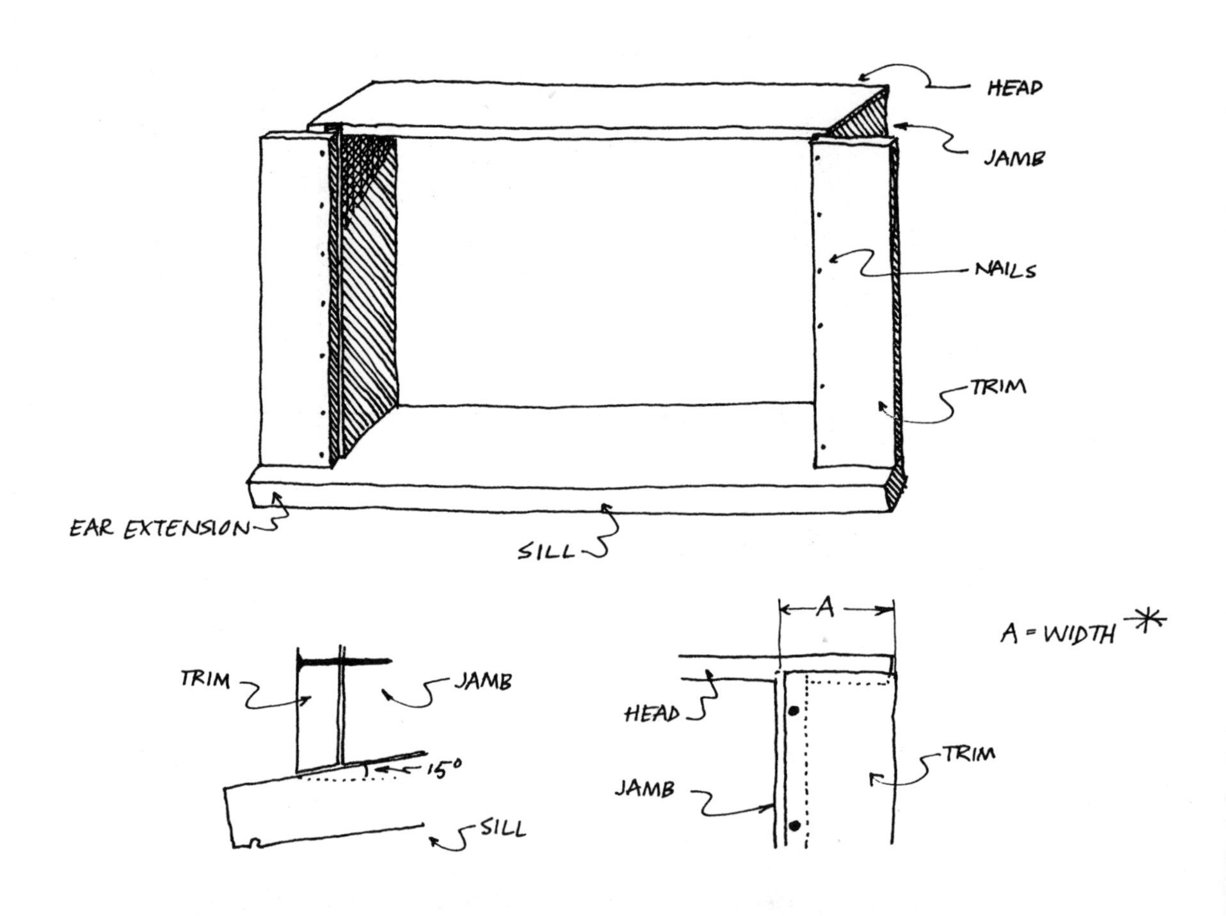

The width * is a matter largely of taste. In older American buildings it is usually (5/4" ×) 6"; in new buildings it is often reduced to (1" ×) 4". The future metric standard may well be (30 mm ×) 100 mm, but this has not yet been established in America. At any rate the trim must be wide enough to reach from jamb to stud and should equal the ear extension of the sill.

After the side pieces of trim have been fixed, the length of the top piece of trim may be accurately measured (from the outside of one side piece to the outside of the other). On the top of this is nailed a piece of drip cap - of the same length.

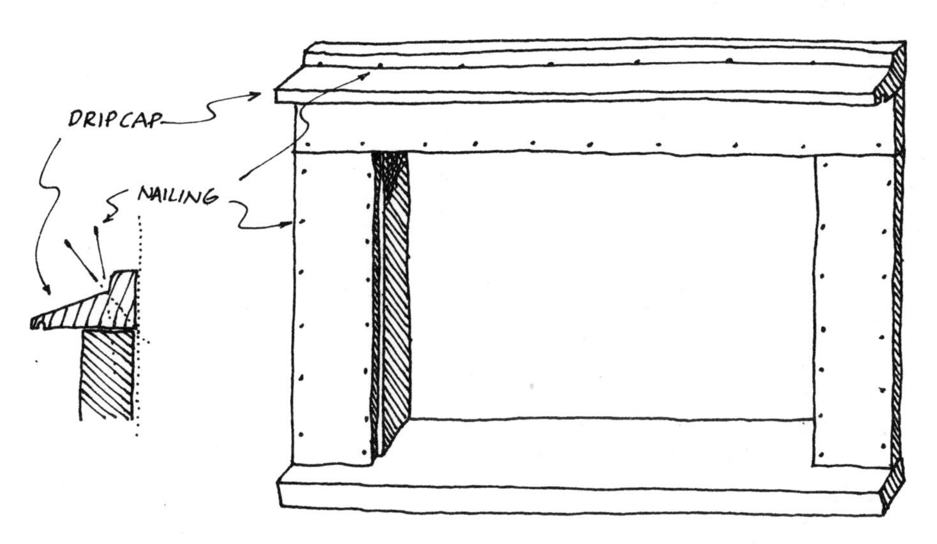

COMPLETED EXTERIOR TRIM (NAILS SHOULD BE SET AND FILLED)

Examination of older buildings, especially colonial style, will reveal the variety of trim details that is possible. What has been here described is the basic technique, which may be adapted to suit the architectural style of your fancy.

As with exterior trim, interior trim is also subject to a wide variety of styles. However, in essence it is the same as the exterior trim, though usually smaller in proportions. There is, of course, no drip cap (though there may be a cabinet head - see the previous chapter) and there are usually two extra pieces not found on the outside: the stool and the apron.

The stool is fitted first, the sides being extended, like the sides of the sill, to be at least as wide as the side pieces of trim.

Then the apron is cut the same length (or 1" (25 mm), shorter if the ends of the stool are finished in a nosing) and nailed firmly under the stool.

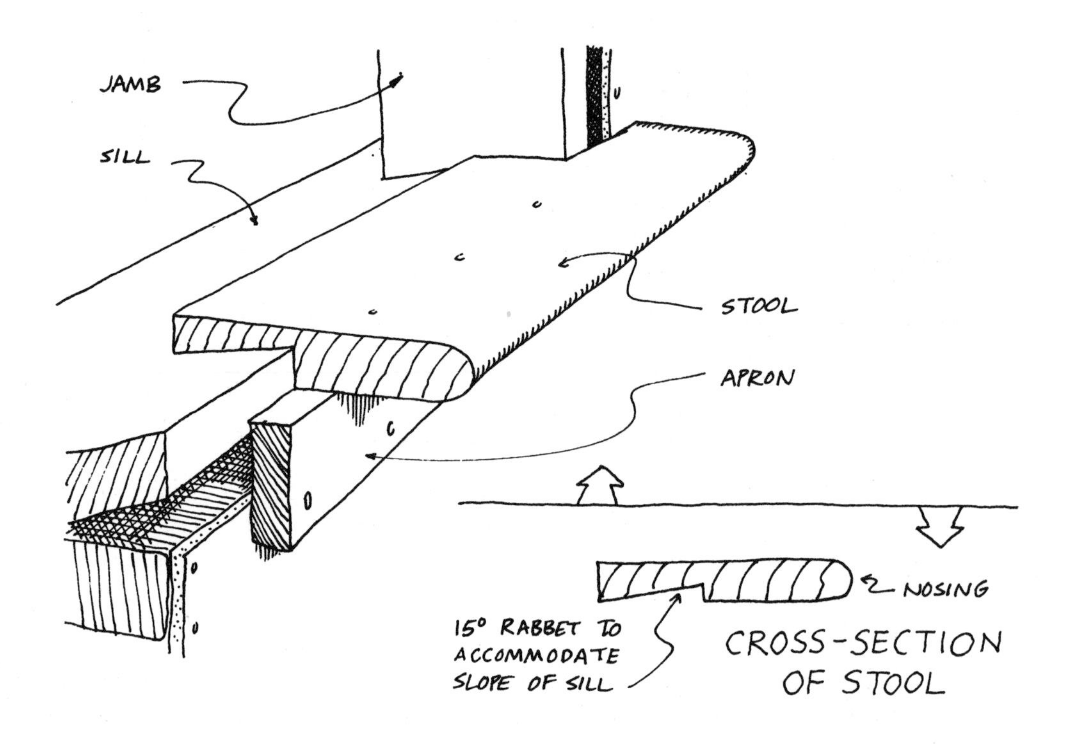

How far the stool extends across the top of the sill can depend on a number of factors, one of the most common being the size and location of the stop – to be installed after the sash has been hung. Another factor influencing the shape of the stool would be the hinging of the sash to open inwards.

After the stool and apron are in place, the side pieces of trim are cut and nailed just as were the outside side trim; and so with the top piece also.

It should be pointed out that such trim was commonly made on the job and often consisted of moulded pieces superimposed. Nowadays the tendency is to use much simpler and smaller trim. As with door casings, trim for windows is readily obtainable in various patterns at most lumberyards.

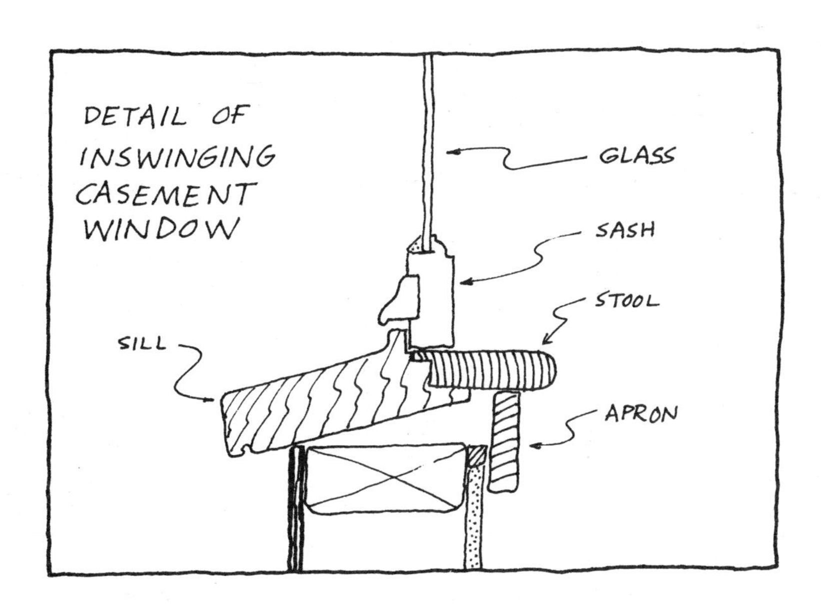

With the frame built, and the trim holding it securely and squarely, the next step is to hang the sash. Remember, the sash should fit into the opening with a little clearance all round, and the bottom should be beveled to match the angle of the sill's slope.

Also, as with doors, the inside edge of the side opposite the hinges should be beveled slightly to allow the window to open, and yet not leave too big a gap.

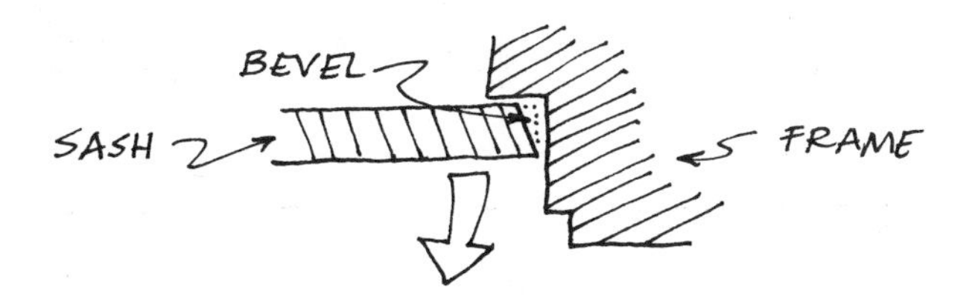

Hinging sashes is very similar to hinging doors on a smaller scale. At the risk of a certain repetition the main points to bear in mind are illustrated below for the sake of completeness.

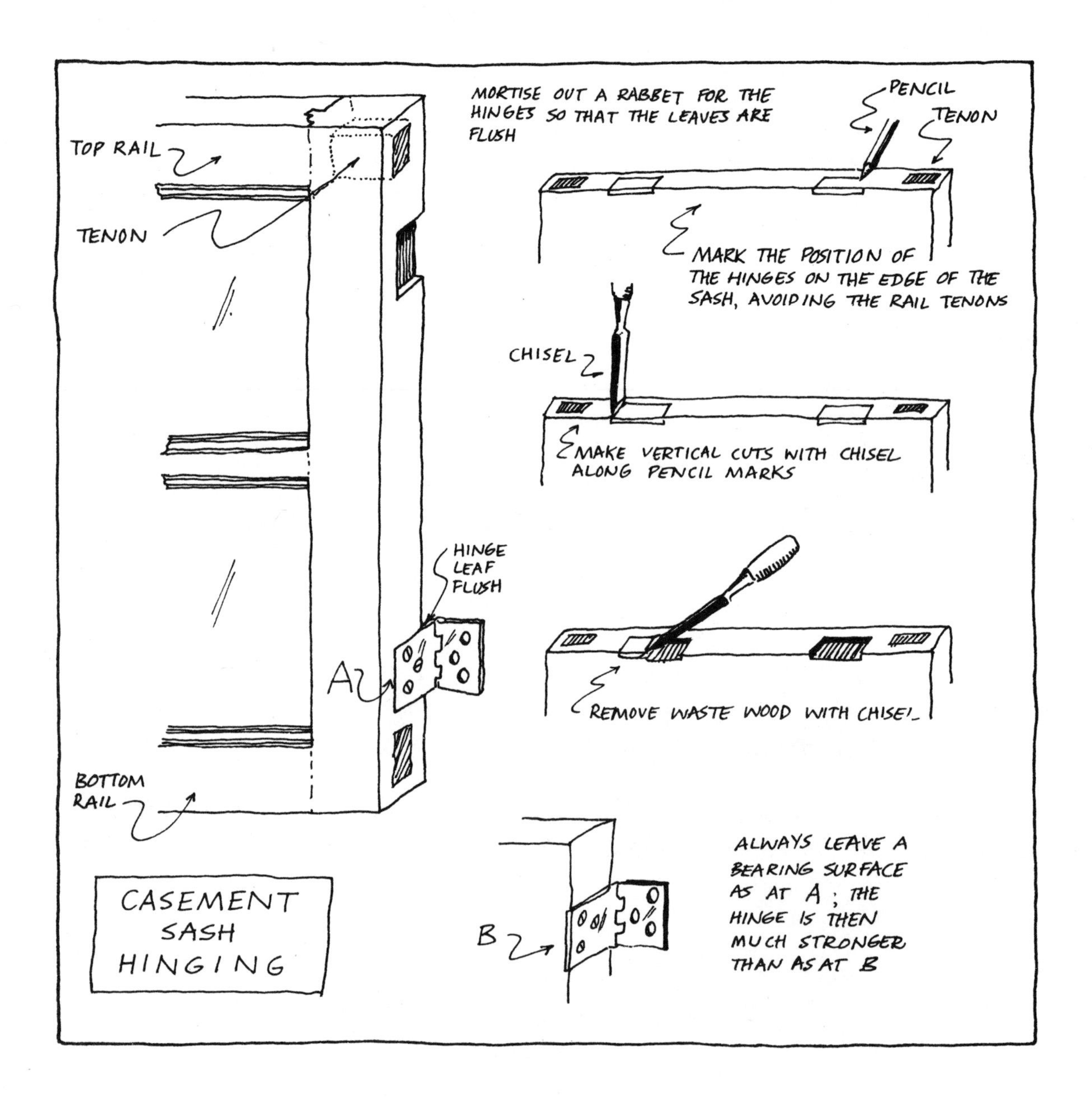
TOP RAIL
TENON
BOTTOM RAIL
HINGE LEAF FLUSH
A
MORTISE OUT A RABBET FOR THE HINGES SO THAT THE LEAVES ARE FLUSH
PENCIL
TENON
MARK THE POSITION OF THE HINGES ON THE EDGE OF THE SASH, AVOIDING THE RAIL TENONS
CHISEL
MAKE VERTICAL CUTS WITH CHISEL ALONG PENCIL MARKS
REMOVE WASTE WOOD WITH CHISEL
B
ALWAYS LEAVE A BEARING SURFACE AS AT A ; THE HINGE IS THEN MUCH STRONGER THAN AS AT B
CASEMENT SASH HINGING

When the hinges are screwed to the sash, hold the sash in the frame and mark the position of the hinges on the frame, the outside of the sash being flush with the outside edge of the frame. Then cut the mortises for the hinges in the frame and screw everything together. To make life considerably easier, use loose pin butt hinges so that each leaf may be affixed separately.

When the sash is hung, the only job that remains is to make the stop. The stop is the strip that runs around the inside of the frame and against which the window closes. If you have built a stool this will usually serve as the bottom stop.

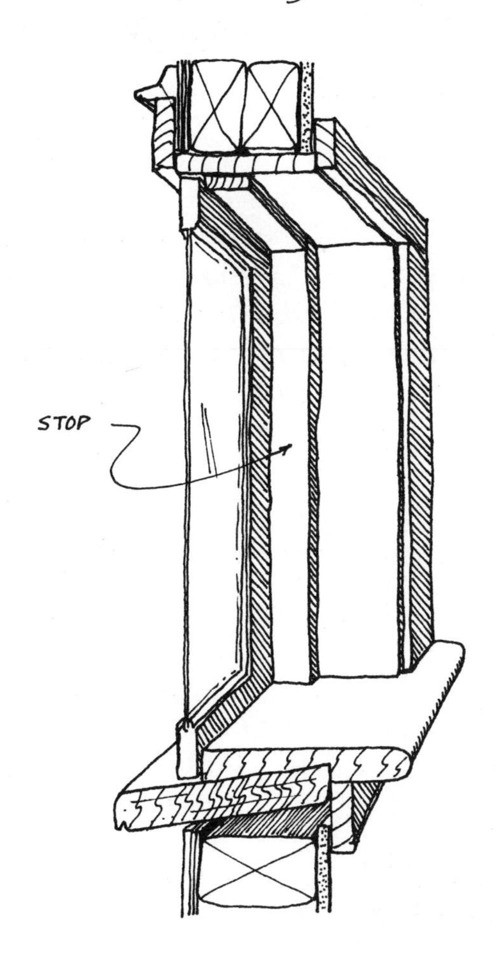

COMPLETED CASEMENT WINDOW

The remaining stop, for the sides and top, may be bought or made, and fitted in the same way as door stop, discussed in the previous chapter.

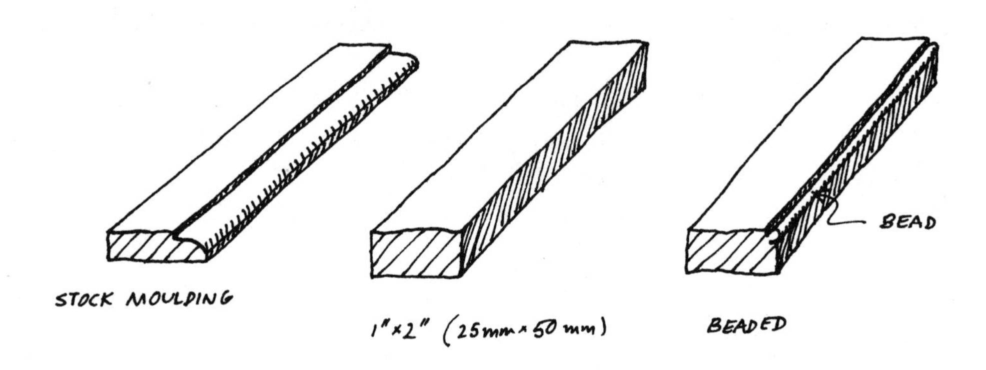

If you intend to fit a storm sash - for insulation - then the stop had best have square sides so that both sashes may close to it. Similarly, when fitting hardware such as handles and casement fasteners remember to make the stop wide enough to allow sufficient space between the two sashes.

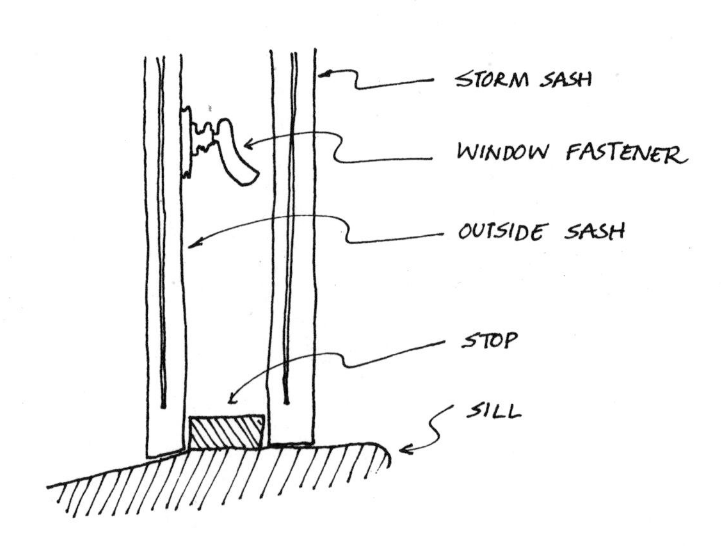

2. DOUBLE HUNG WINDOWS

Double hung windows are more complicated than casement windows, but their basic construction is as follows.

The jambs (in the best type of construction) are rabbeted into two inner casing members which form a box with the rough framing in which the sash weights move up and down.

The jambs, in this case, are properly known as the yoke and the pulley stiles, as shown below.

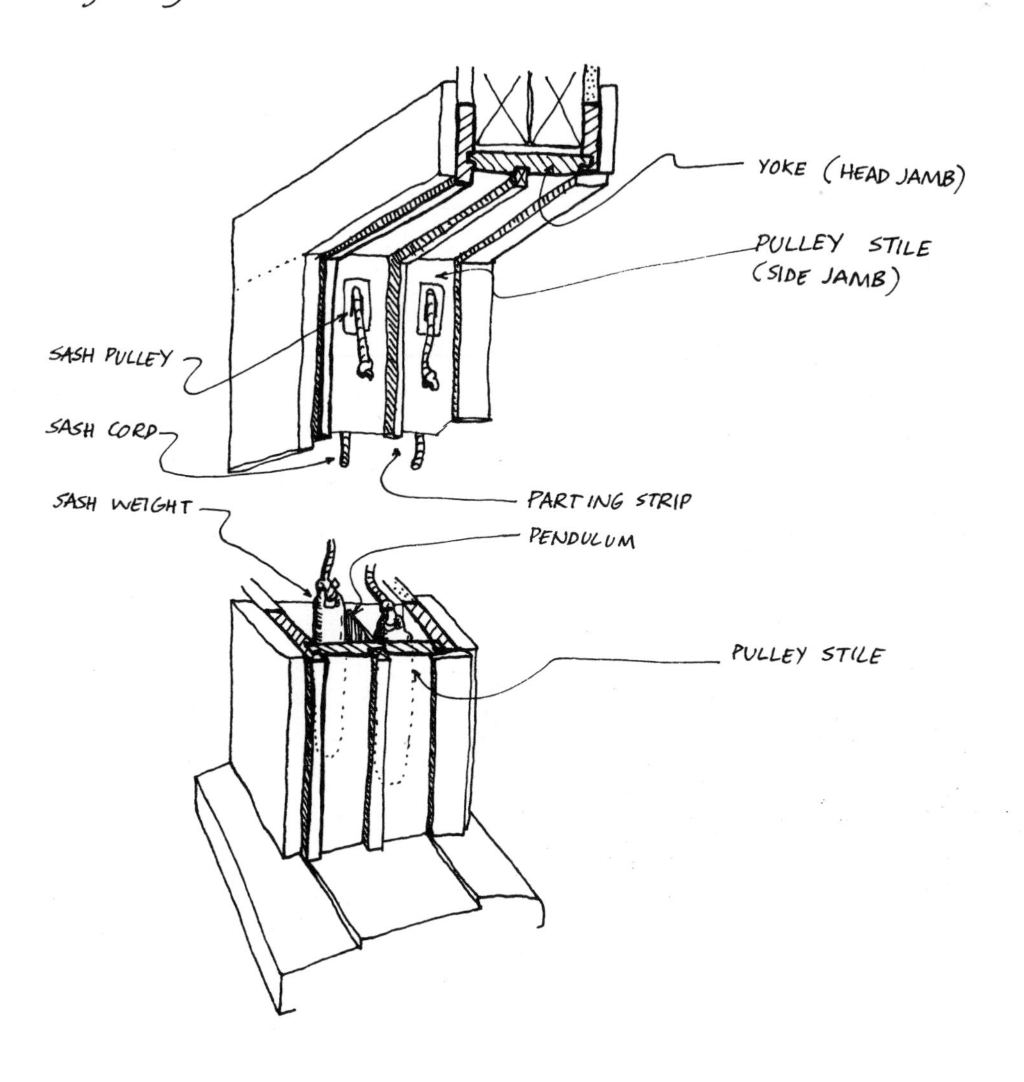

The yoke and the pulley stiles, which are also referred to as the inner or rough casing, provide a nailing surface to the rough opening. The inside rough casing should lap the rough opening framing members about 1" (**25 mm**). The outside rough casing forms a blind stop for the sheathing.

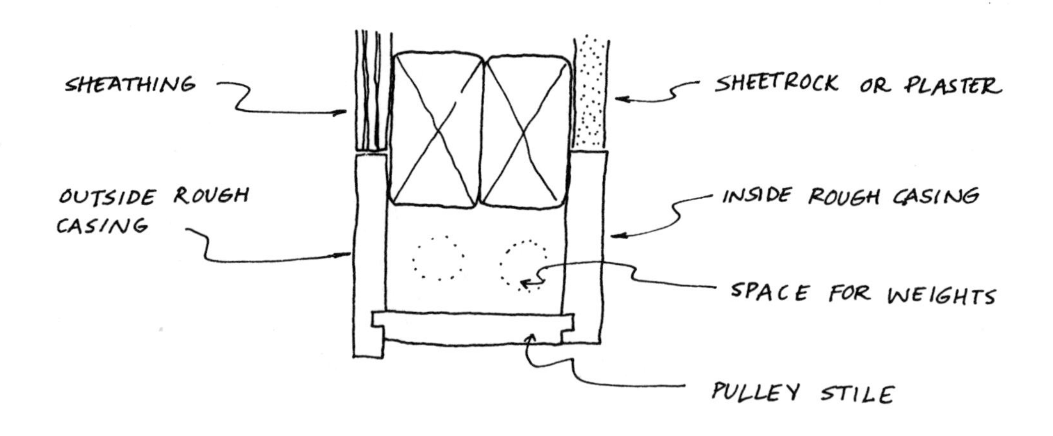

The space between the rough opening and the pulley stile forms the box for the sash weights which counterbalance the sash. In better class construction this space is divided by a vertical piece of wood known as the pendulum. At the bottom of the stiles is usually a removable panel through which the weights may be retrieved should the sash cord break.

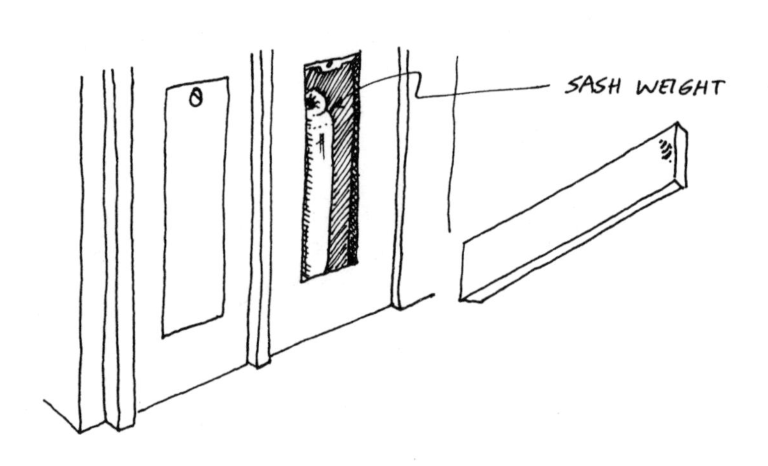

The faces of the yoke and stiles have a piece of wood dadoed into them, called a parting strip. This is removable in order that the upper sash may be taken out when necessary. The strip forms a center guide for the upper and lower sash, while the outer rough casing projects slightly beyond the stiles and yoke and so forms the outer guide. The inner guide is formed of regular window stop, which is also removable in order to allow the lower sash to be taken out from time to time.

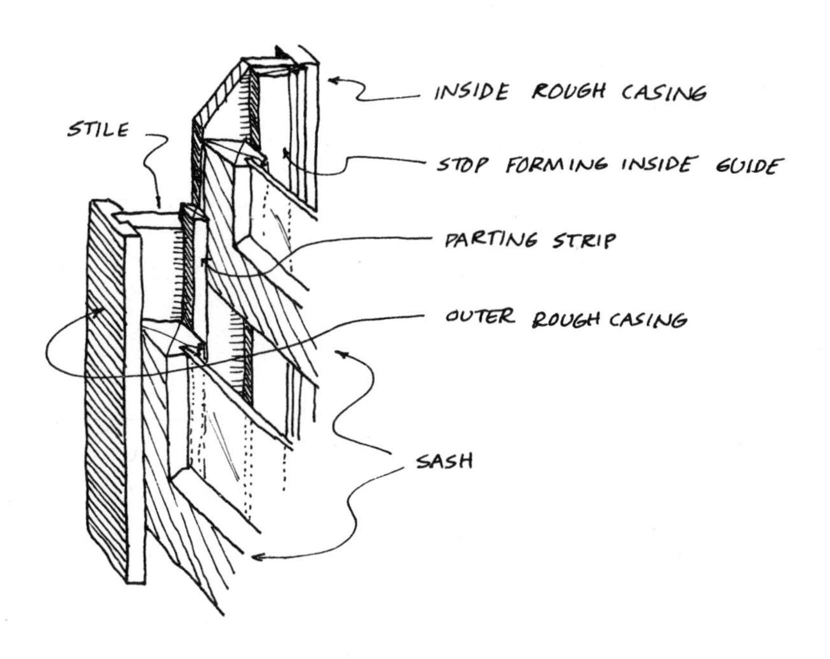

At the top of the stiles two pulleys on each side (two for each sash) are mortised in flush with the stile.

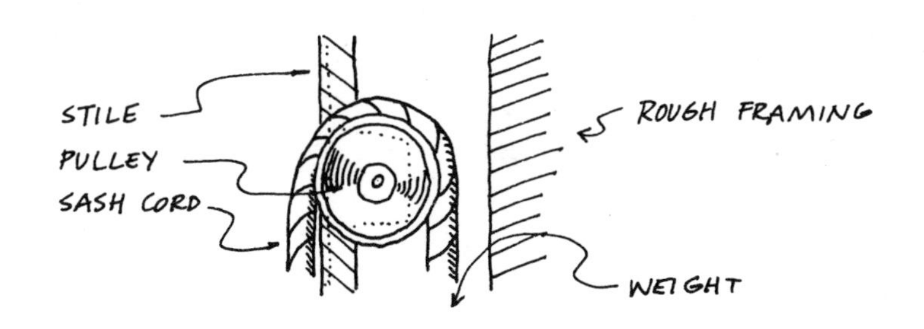

The sill is an integral part of the frame and must slant downwards and outwards in order to shed water. It should have small steps, one at the point where the lower sash rests on the sill, and another near the outer edge to form a seat for window screens and storm sash. These steps prevent water, which drips on the sill, from being blown up under the sash.

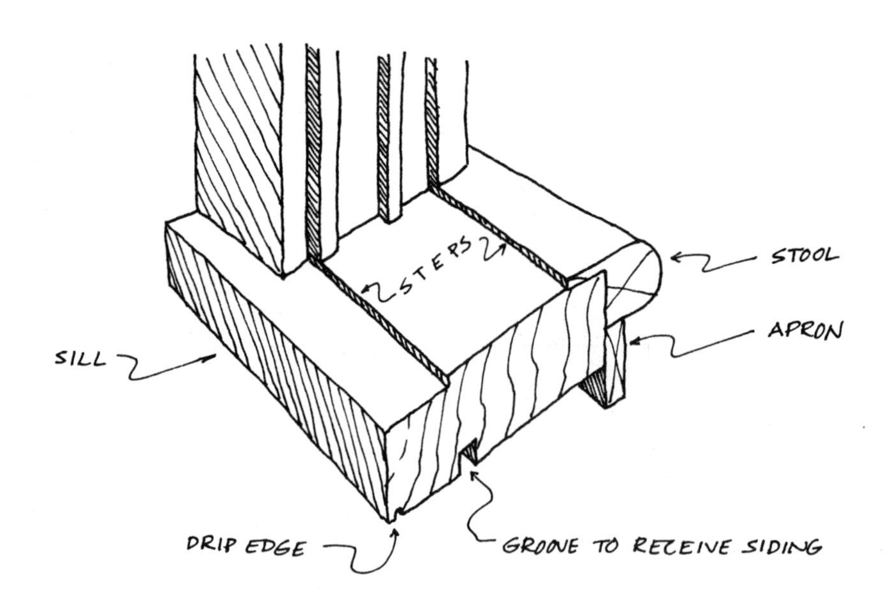

The inside of the sill is finished as with a casement window – with stool and apron and trim. Likewise, the outside is finished with all the exterior trim, including drip cap.

Screens may be fitted as required, but it is always a good idea to install weatherstripping and storm windows in order to reduce heat loss when the building is being heated; the use of double glazing will also help keep the building cool during hot weather, and so reduce air-conditioning bills.

Chapter Three

Walls and Ceilings

With the exception of tiled and suspended ceilings, what may be used for walls may be used for ceilings too, and very often the same material is used for both in the same room, which is why walls and ceilings are here dealt with together in the same chapter.

In the old days there were really only two things you could do to finish the interior of a house: plaster it or panel it. With the development of new materials and new applications the list has expanded and now includes:

1. PLASTER
2. SHEETROCK
3. WOOD PANELING
4. PLYWOOD PANELING
5. TILED AND SUSPENDED CEILINGS

1. PLASTER

Barely fifty years ago plastering, to finish the interior walls and ceilings of a house, was the rule rather than the exception, but nowadays this is reversed, and although it is still possible to find an experienced plasterer, in most communities the majority of houses are finished with one or other of the "dry wall" techniques discussed further on in this chapter.

Nevertheless, I shall outline the basic preparatory procedure, since in many ways, particularly if cost is not a consideration, plaster is preferable.

Plaster requires some kind of base to which it is applied. There are three main types :

a. WOOD LATH
b. METAL LATH
c. ROCK LATH (GYPSUM BOARD)

a. WOOD LATH This is almost completely obsolete now but used to consist, in very old buildings, of green oak cut accordion fashion and spread apart on the framing.

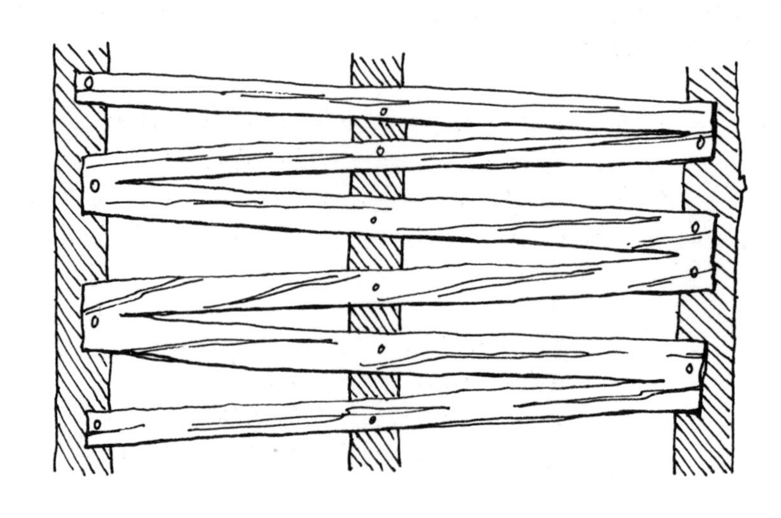

A subsequent and better development was the use of thin strips of white pine, usually 1½" × 3/8" (40 mm × 10 mm) in 4' (1.2 m) lengths (which allowed the lathing to be nailed economically to the framing, which was either 12" (305 mm) or 16" (406 mm) apart. The lathing is applied horizontally about 3/8" (10 mm) apart. If the lath is dry it should be soaked in water before being applied to prevent it from drawing the water from the plaster.

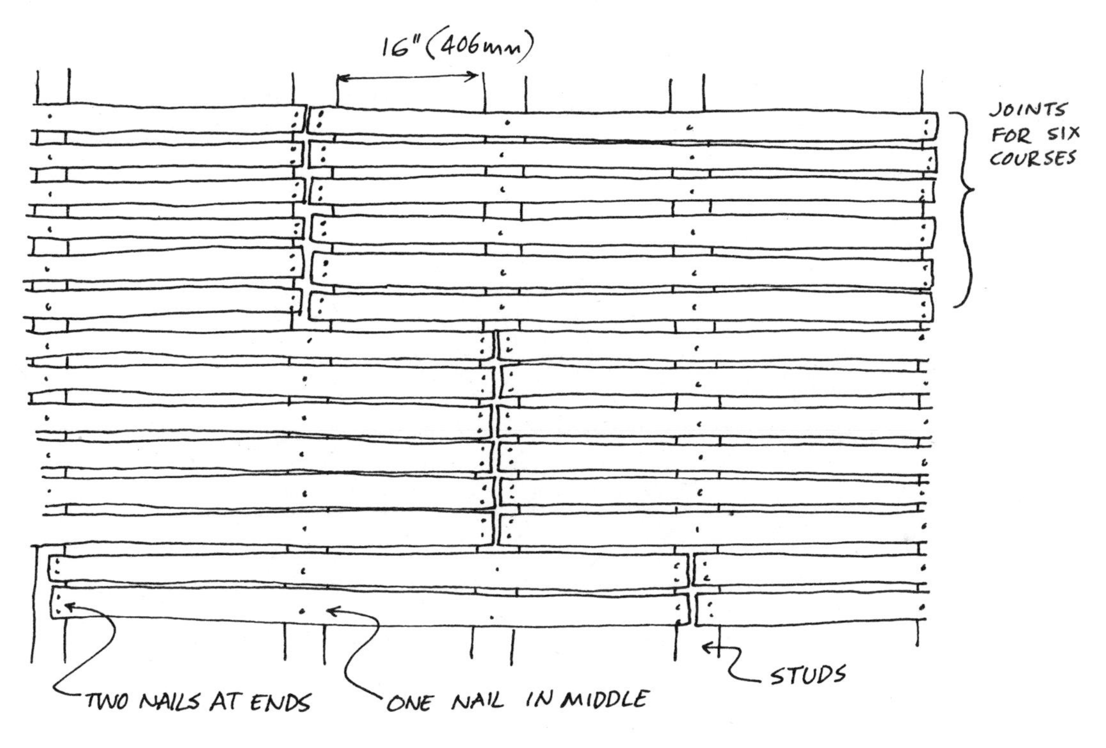

LATHING PLAN

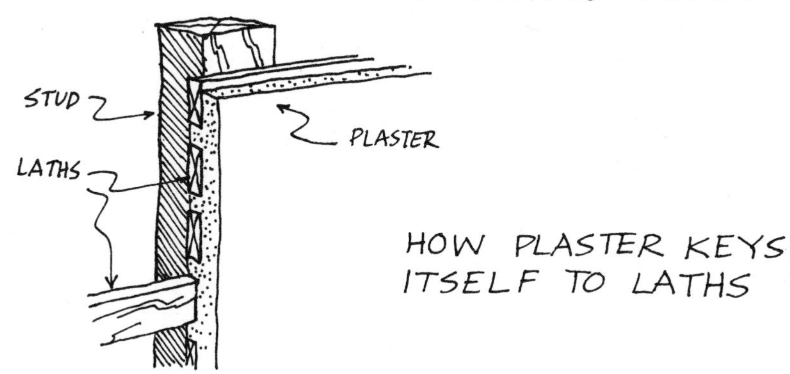

HOW PLASTER KEYS ITSELF TO LATHS

b. METAL LATH Expanded metal lath is much more common as a plaster base than wood lath and is much quicker to apply as it usually comes in sheets measuring 27" × 96" (686 mm × 2438 mm). It consists of sheet metal that has been slit and expanded to form innumerable openings into which the plaster is forced, thus keying it firmly to the metal base – which in turn is nailed securely to the framing. It is usually painted or galvanized to resist rusting. It is easily cut with tin snips and the different pieces wired together every 12" (305 mm) or so.

There was once a great variety of patterns available, but now the commonest is the herringbone pattern illustrated below.

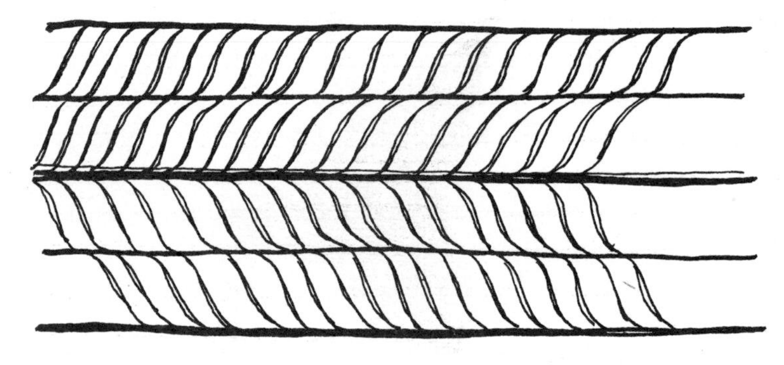

SECTION OF HERRINGBONE METAL LATH

The longitudinal ribs are set at an angle of 45° and the cross strands are flattened. The ribs act as shelves and hold the plaster. The cross strands curl the plaster behind the lath, completely covering it.

C. ROCK LATH The most common plaster base in use today is rock or gypsum lath. This comes in sheets measuring 16" x 48" (406 mm x 1219 mm) and consists of two paper faces with a gypsum filler, and sometimes with a foil backing that serves as a vapor barrier. If the framing to which it is attached is the usual 16" (406 mm) on center, then 3/8" (10 mm) rock lath is sufficient. If the framing is 2' (50 mm) on center, then 1/2" (13 mm) should be used.

A better type of rock lath is perforated, which improves the bond between lath and plaster and which increases the time the plaster stays intact in a fire, for which reason many local building codes specify this kind only.

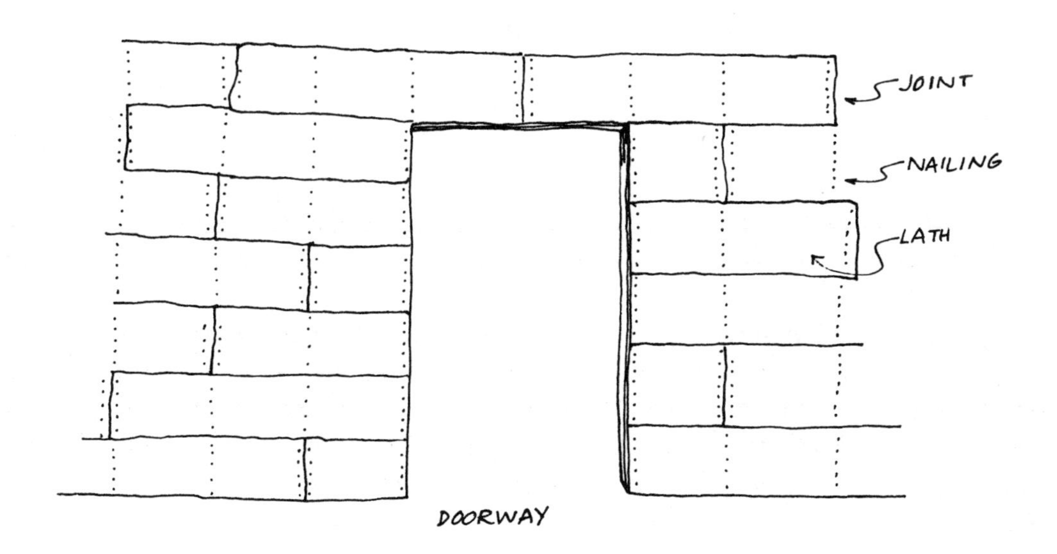

HOW TO BREAK THE JOINTS BETWEEN COURSES

APPLICATION

Before applying the rock lath make sure you have adequate nailing surfaces in all corners and around all openings.

The lath is applied horizontally with the joints broken as shown on the previous page. Vertical joints should all be over studs, and sheets should not join over door and window jambs.

Nailing should be 4" (100 mm) to 5" (127 mm) on center, and 1½" (38 mm) blued and ringed lathing nails should be used.

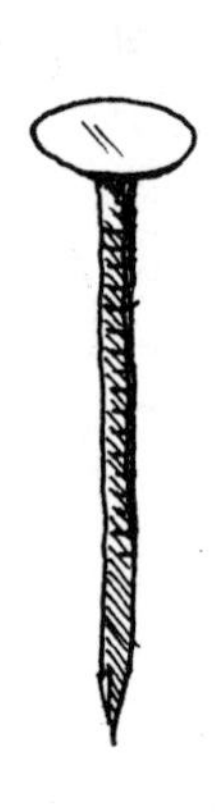

It is often easier to fix the plaster grounds before fixing the base – although with rock lath the grounds are often fixed afterwards.

The plaster grounds are strips of wood the same thickness as the base and plaster (if applied before the base) which are attached to the framing around window and door openings, and along the floor line. They serve as a leveling surface and a plaster stop for the plaster, and along the floor line they also provide a nailing surface for the baseboard.

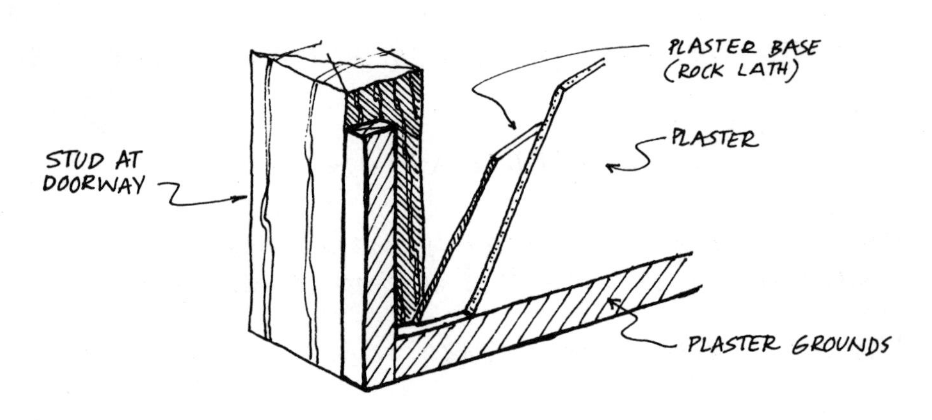

Strip grounds are nailed to the INSIDE *of the various rough openings, and are removed after plastering.*

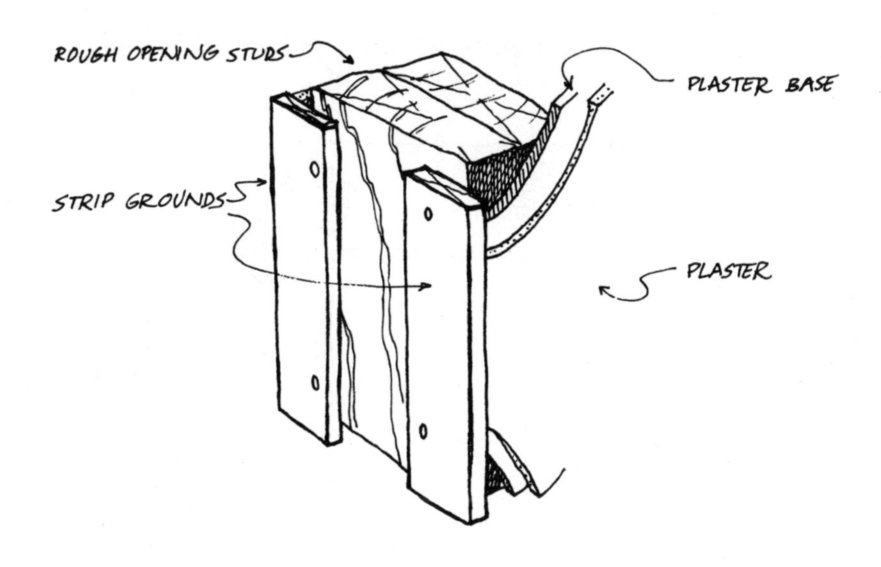

If the window and door frames are already in place, as is very often the case, then these serve as grounds – which means you must have calculated the width of the plaster and its base into the width of the door jamb. If the frames are not wide enough, you will have to fix grounds and fur the frames out before attaching the trim.

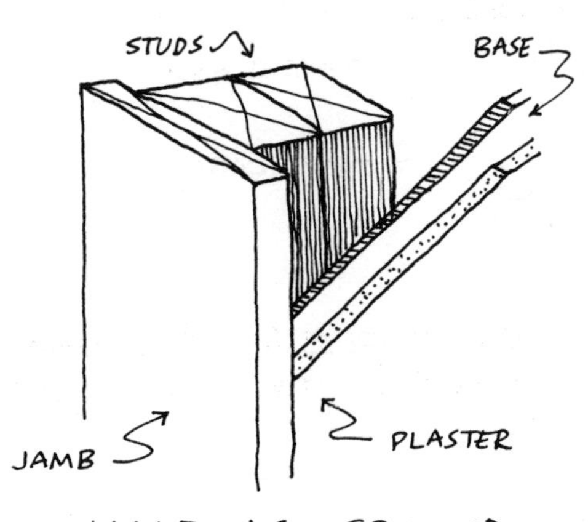

JAMB AS GROUND

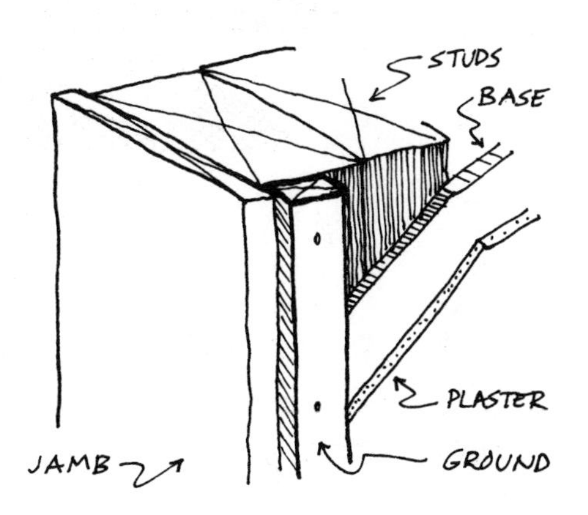

NARROW JAMB REQUIRING EXTRA GROUND

To avoid cracks in the plaster caused by shrinkage of the wood framing as it dries out (in a new construction), expanded metal lath is used as reinforcement in the following places:

1. Over door and window openings as shown below.

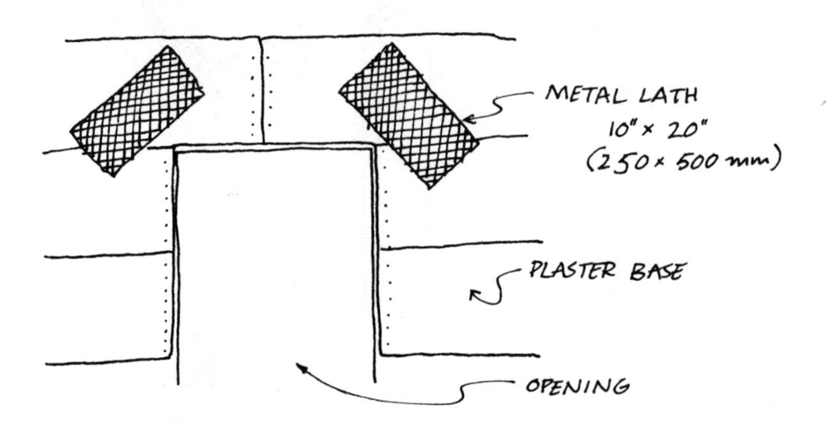

2. Under flush ceiling beams as shown below.

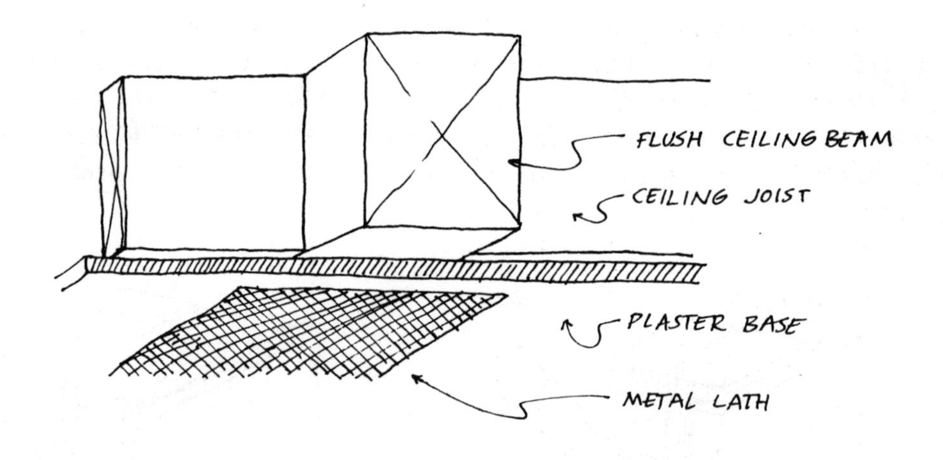

3. At all exterior corners, as shown opposite, it is usual to use a metal corner bead which is plastered over and so forms a rigid corner.

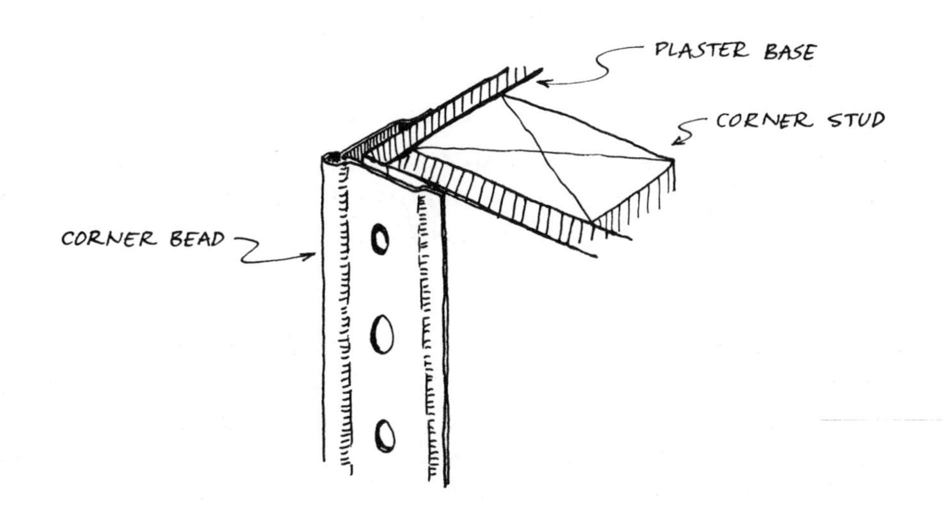

The plaster, which consists of sand, lime or special plaster, and water, should be at least ½" (13 mm) thick when finished.

Over wood and metal lath it is usual to apply three coats of plaster. Over rock lath the first two coats are combined and only two coats are necessary.

In three-coat work, the first coat is called the scratch coat because it is scratched before it is completely dry to ensure a good bond for the second coat, which is called the brown or leveling coat. When using rock lath as a base – which is presumably level when applied – the first two coats are combined.

The finish coat in both cases may be either rough (sand-float finish – achieved by mixing lime with sand) or smooth (putty finish – with no sand).

The mixing and application requires considerable skill; you should not expect good results the first time. There are many considerations to be borne in mind with regard to speed of drying and temperature, etc., and plastering itself is best left to a professional plasterer.

2. SHEETROCK

Sheetrock, variously referred to as drywall or gypsum wallboard, is by far the commonest material used today for walls and ceilings, and though not properly a carpenter's skill, its u is not so hard that a layman, with care, might not expect good results the first time.

It comes in sheets 4' (1.2m) wide by 8' (2.4m) or longer in length, and in three thicknesses (3/8" (10mm), 1/2" (13mm), 5/8" (16mm)) for use with differently spaced framing – the wider the studs are apart, the thicker should be the sheetrock. The edges of the sheets are tapered on the face side so that the joints may be filled and taped smoothly.

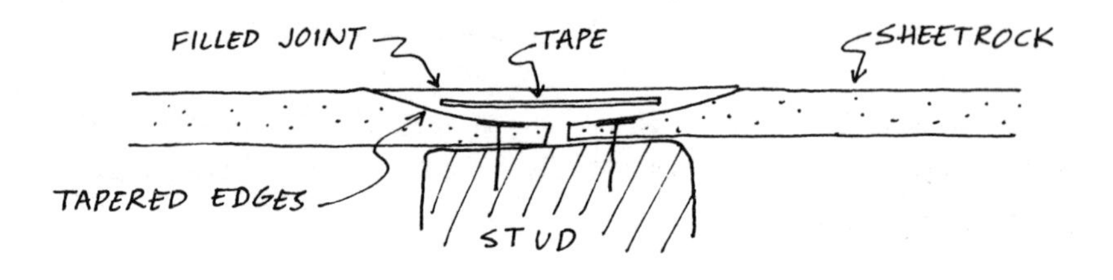

Professional sheetrockers use power-driven screws and nail guns, but good results may also be obtained with a regular hammer, using blued and ringed sheetrock nails like the one shown below.

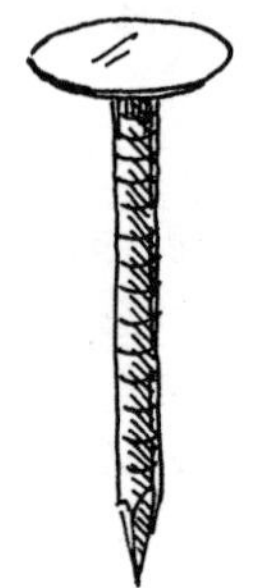

Since drying of the framing may cause the nails holding the sheetrock to pop out a bit, it is best to use these ringed nails (it is even better to use the screws) and to allow the framing to dry out as much as possible before sheetrocking.

The sheets may be fixed vertically or horizontally – it is stronger horizontally, and also creates fewer joints this way – but each situation will dictate the best arrangement. It is often easier to start from the corners when covering walls, and if you are doing the ceilings too, do them first so that the ladders or scaffolding are out of the wall of the finished walls. Do not butt the sheets tightly together; leave 1/8" (3 mm) space between sheets to allow for minor expansion changes.

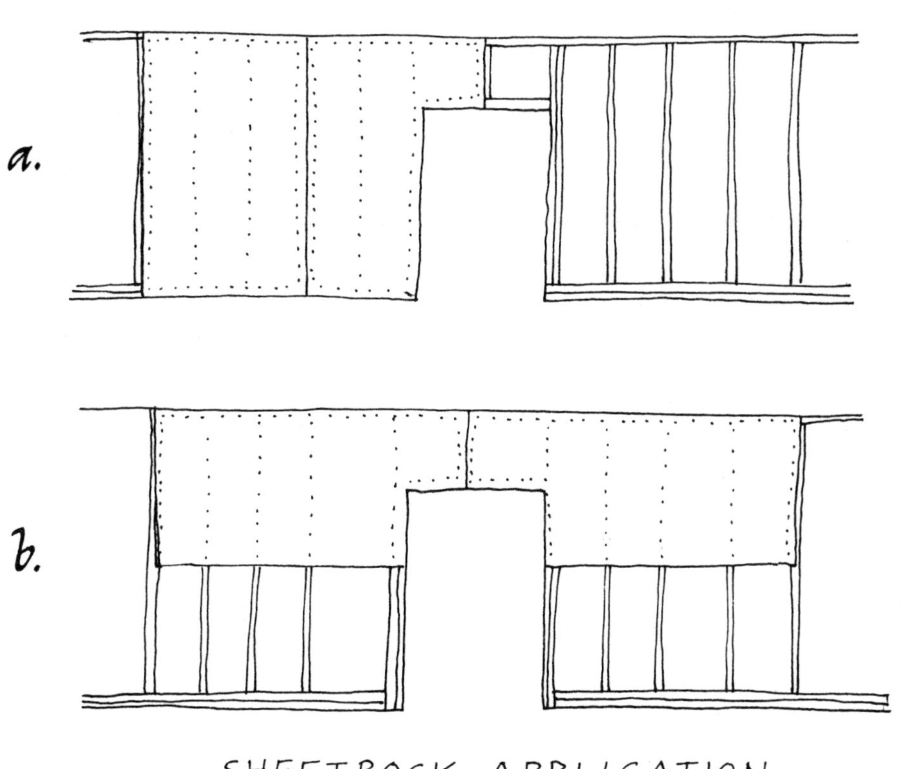

SHEETROCK APPLICATION

a. VERTICAL
b. HORIZONTAL

Sheets are normally delivered in pairs, with the finish surfaces facing each other, for protection. They are heavy, especially the longer and thicker sizes, so if you have a lot of sheetrocking to do, try and have the material delivered where it will be needed - not all in one place, which might be more convenient for the delivery man.

To cut the sheets to size, or to make holes for electrical switches and outlets, score the face side first with a sheetrock knife, in order not to tear the face paper.

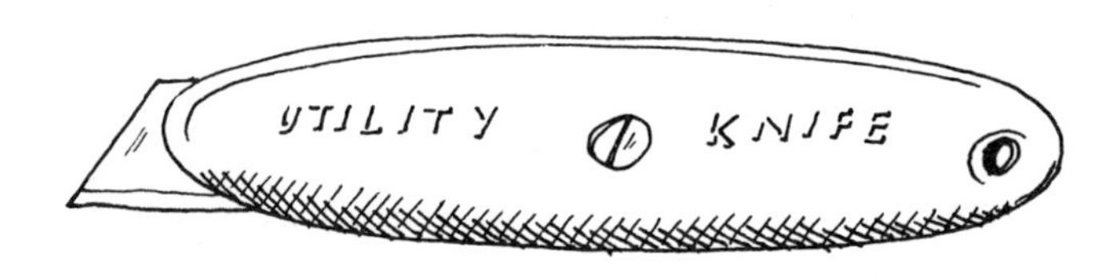

Then fold the sheetrock back on itself and score again in the crease. Finally, fold it the other way and it will break off.

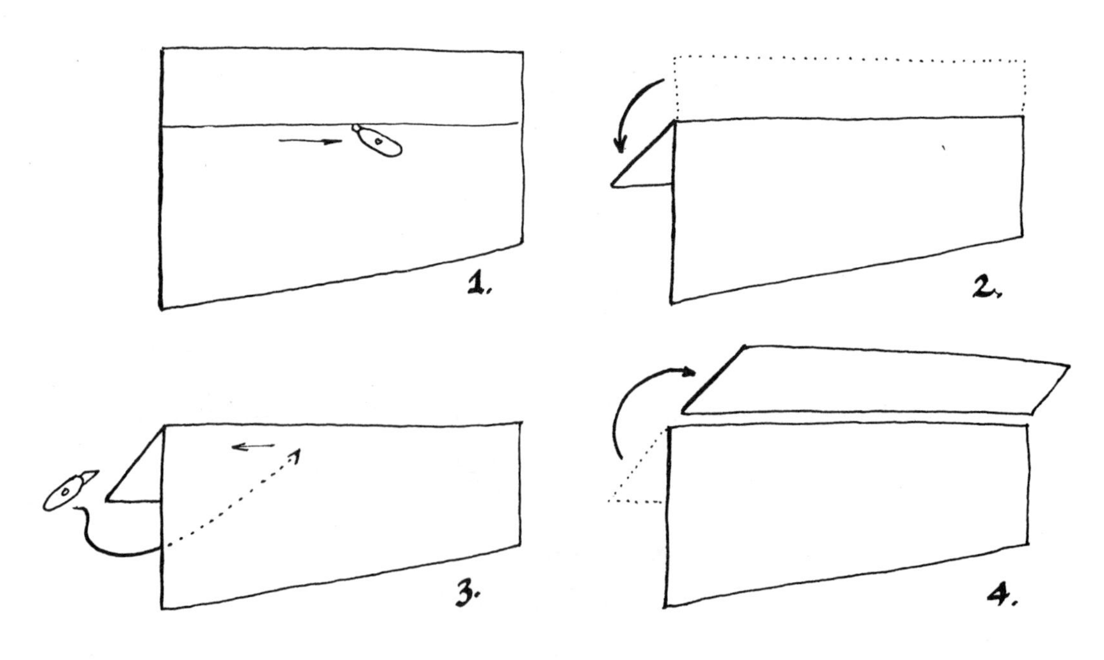

Of course this only works if you can fold an entire flap. If you want only to cut part of the way, you must use a saw or, less messy but more effort, keep scoring with the knife until you are all the way through. Several light passes are easier than one heavy pass. Also, use a straight edge to guide the knife, or at least snap a chalk line to ensure straight cuts.

It doesn't matter so much if you tear the backing, but measure carefully – especially the position of holes – and after scoring with the sheetrock knife, use a sheetrock saw for cutting the pieces out.

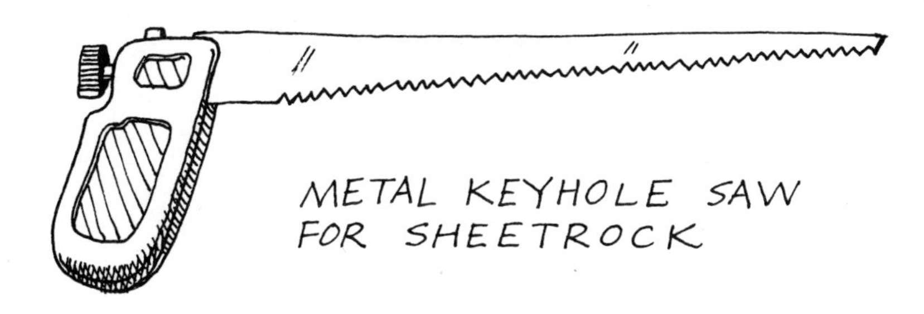

METAL KEYHOLE SAW FOR SHEETROCK

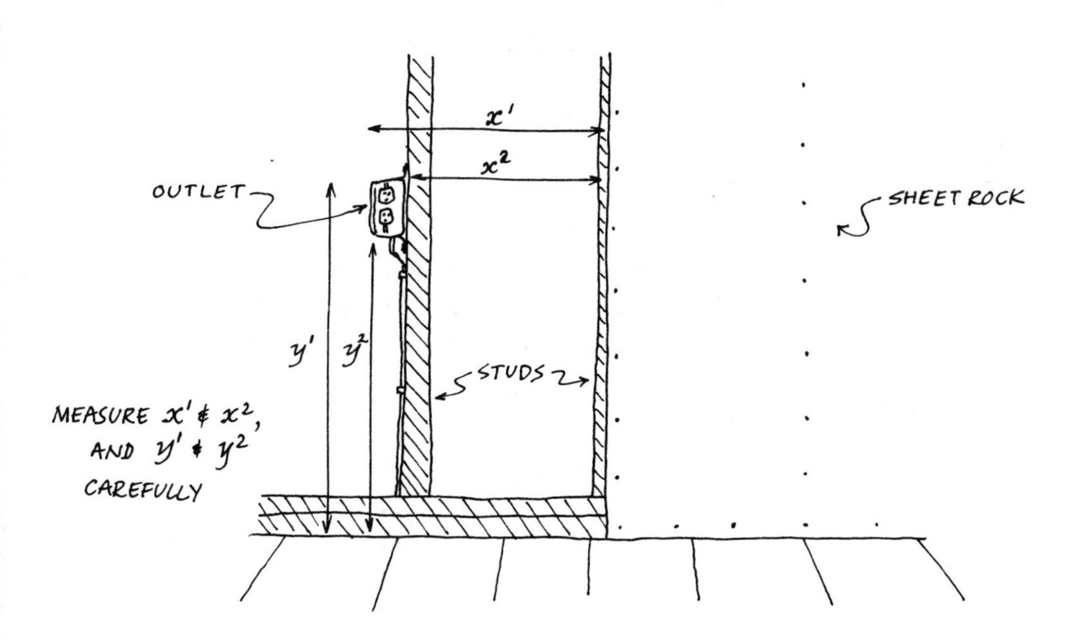

HOW TO MEASURE FOR HOLES

Nailing should be every **6" (150 mm)** or so on studs and around all edges. Alternatively you may space the nails **12" (300 mm)** apart and then nail a second series of nails alongside the first nails (say **2" (50 mm)** apart) – this helps get and keep the sheetrock tight to the studs.

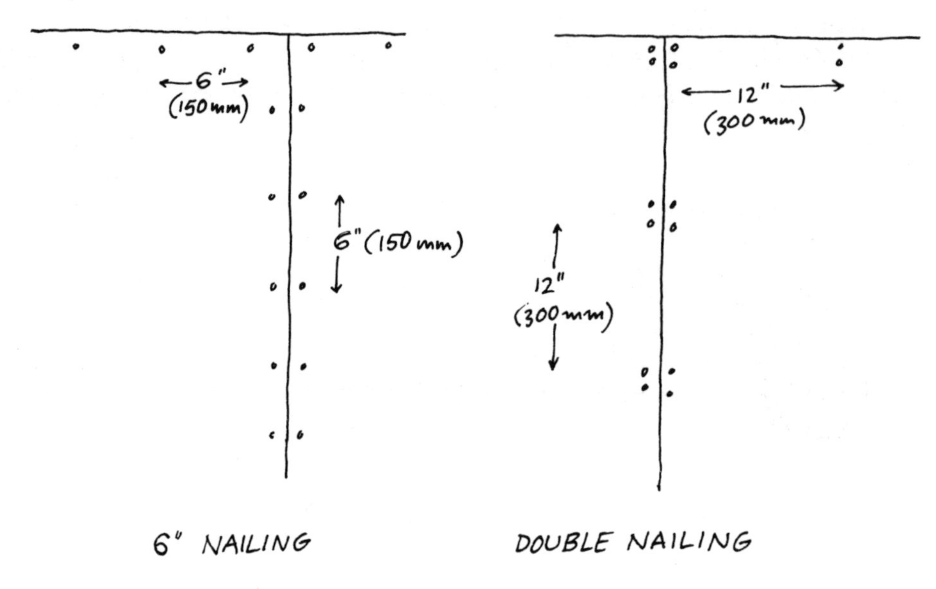

SHEETROCK NAILING

The nails should be hammered in so that a slight dimple is left in the sheetrock - but avoid breaking the paper - in order that the depression may be filled smoothly and the nail head hidden.

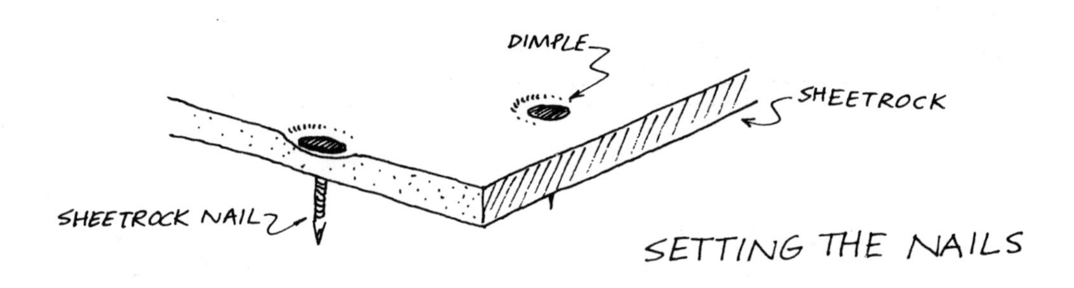

SETTING THE NAILS

When all the area to be covered is sheetrocked, the "taping" begins. Like plastering, this is often something that many people, including professional builders, leave to "tapers" who do nothing else, and who are consequently very good and very fast. If you have a whole house to do, it will be quicker, cheaper, and better if done by professional tapers, but if you have only a small area, you may as well do it yourself.

All that is really needed is patience and perseverance. Firstly all nail heads are covered, using joint compound,

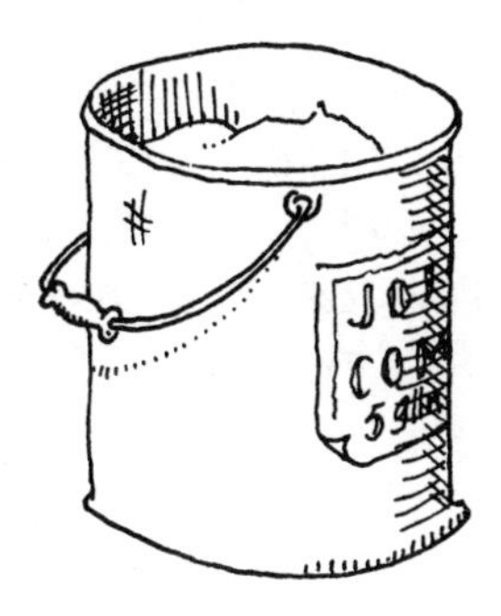

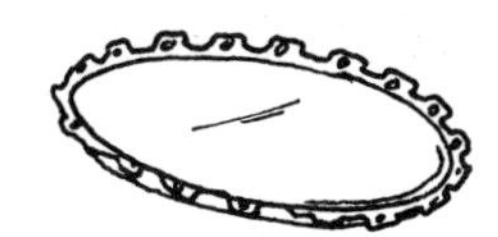

which is most economically bought in **5** gallon (**19** liter) drums, but which may be bought in amounts as small as **1** pint (½ l.). There are various brands of joint compound which go under different brand names. Avoid those with asbestos in them; asbestos is harmful if you inhale the dust when sanding.

To cover the nail heads, use a putty or spackle knife; anything from **2"** (**50mm**) to **6"** (**152mm**) is good.

SPACKLE KNIFE

When all the nail heads are covered, lay in a bedding coat of joint compound along all joints, and in this lay the tape – which comes in rolls; buy it with the compound.

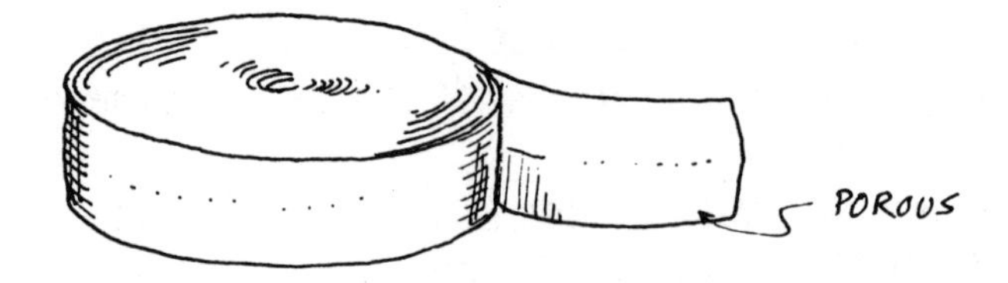

For inside corners, fold the tape first and then lay it in. When the tape is in the wet compound, smooth it down with a knife broader than the tape, taking care to maintain an even pressure, but not pressing too hard or the tape will slide out of the compound and tear.

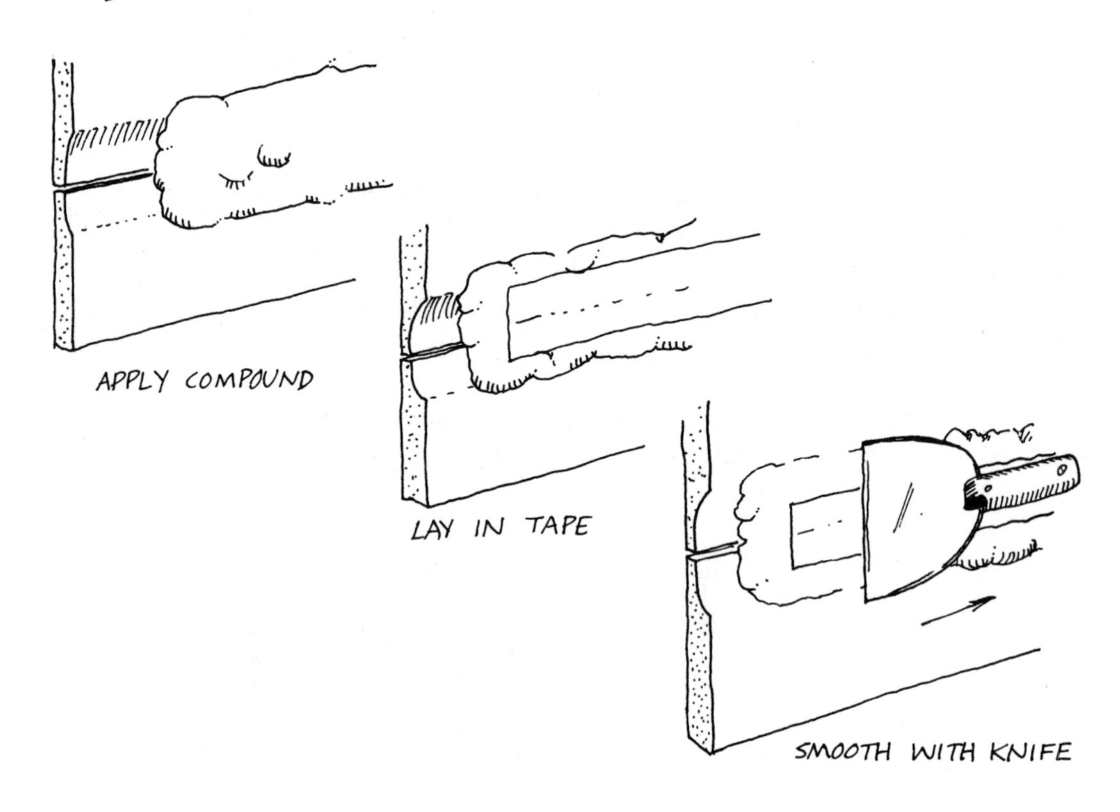

For working in corners you may prefer to use a special "inside corner trowel."

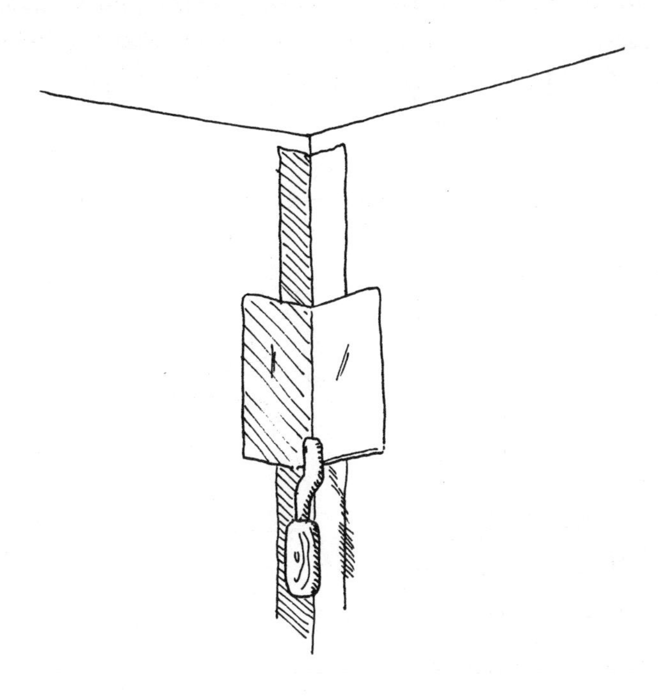

For all exposed corners you should use an appropriate metal trim, which is simply nailed onto the sheetrock and compounded over as if it were tape.

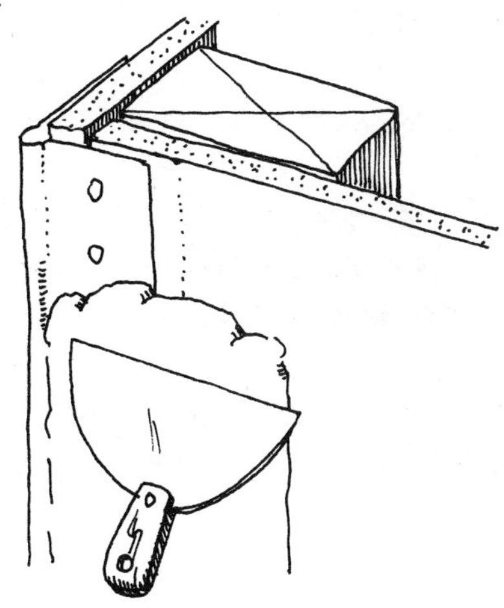

After the bedding coat and the tape are dry, usually twenty-four hours, apply a wider, tape-covering coat and wait for that to dry. This should dry a little faster, say overnight, since you should have used less compound. Finally, apply an even wider, trowel-finished coat, using a very wide knife.

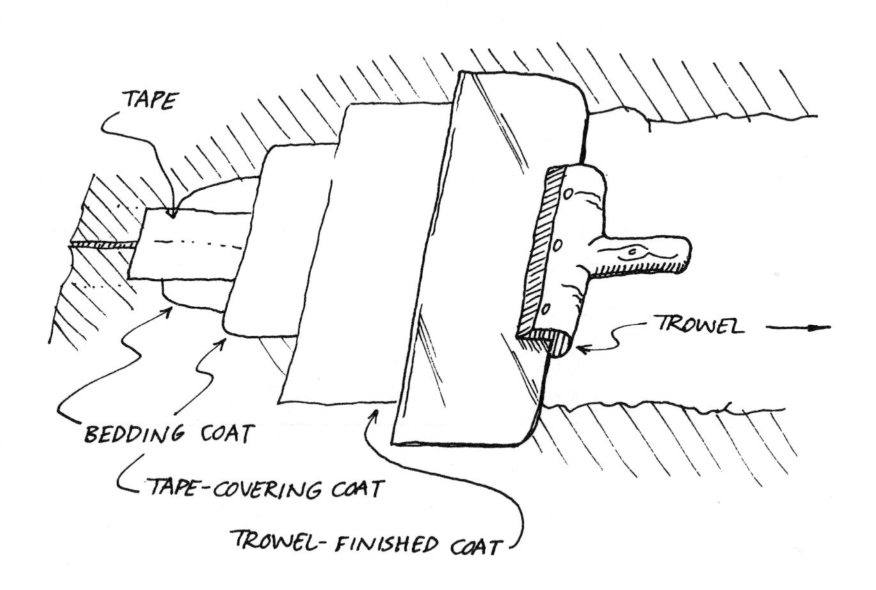

At every coating go over the exposed nail heads again, and, following the tape-covering coat and finish coat, wet-sand, using a damp sponge to avoid possibly harmful dust, to eliminate any bumps or ridges. The more careful you are when applying the compound, the less sanding you will have to do.

When taping inside corners, fold the tape well before putting it in the joint, for it could shrink when dry and end up giving you a round corner.

Lastly, don't bother to tape corners that will be hidden by moulding, such as ceiling and wall joints!

3. WOOD PANELING

Wood paneling, which for so long was the only alternative to plastering, has similarly given way to a preponderance of plywood in various shapes and forms in all but the very best work.

It was developed as a way to cover large areas with only small boards so that nothing would warp and allowance could be made for any shrinkage or expansion. Areas paneled were commonly around fireplaces and the lower portion of living room walls - which was referred to as wainscot.

"Wainscot" meant originally the wooden sides of a wain or wagon, and then referred to a type of imported (into England) Baltic or Russian oak, much used for this kind of work.

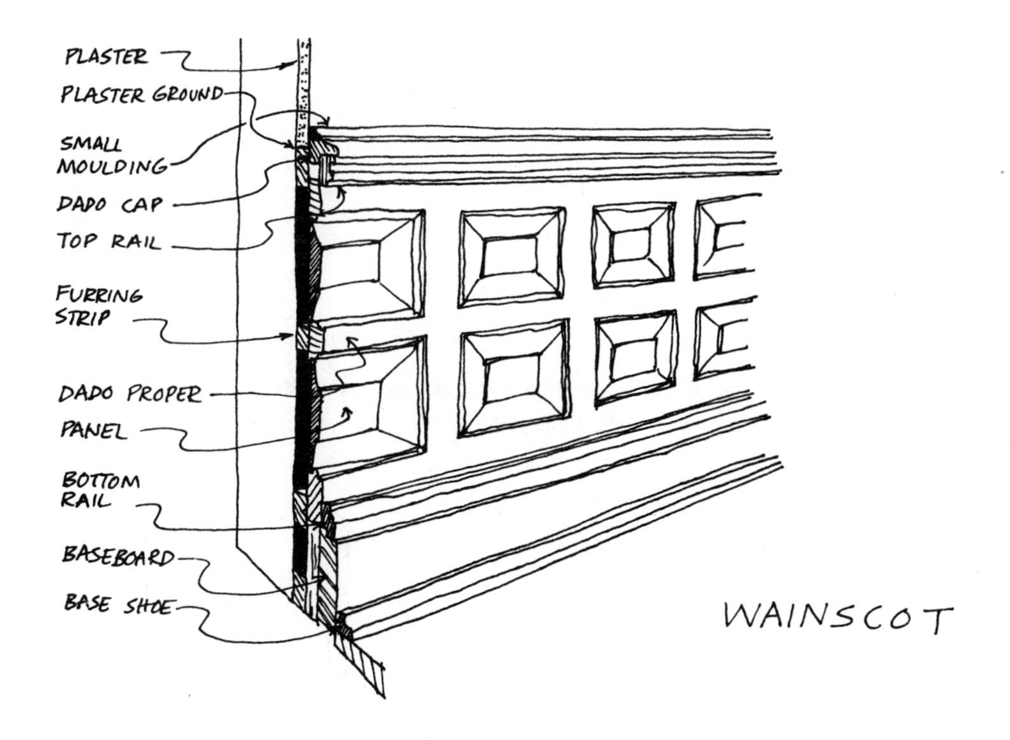

A much simpler form of wood paneling is the use of planks or small boards applied horizontally or vertically (or in any other direction if you will).

The boards may be simply square-edged plain boards, but for good work they are usually matched, that is, tongued and grooved and well-seasoned. If you are using square-edged boards and some shrinking is possible (using green rough-sawn lumber is very inexpensive and popular in some areas, and this shrinks a lot), put up black building paper or felt before you fix the boards. This doesn't stop the shrinking, but it makes the subsequent cracks less obvious.

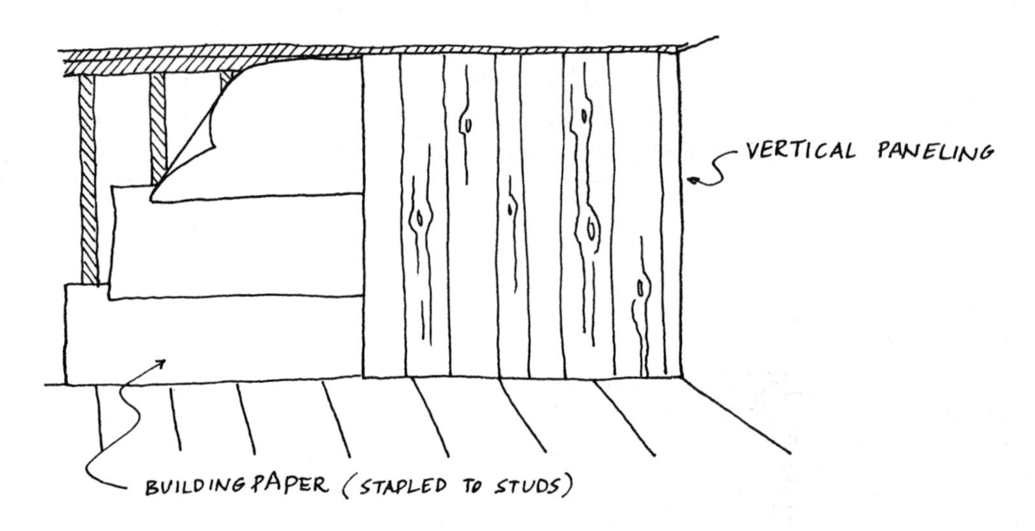

The most important thing with this kind of paneling is to ensure that you have adequate nailing surfaces – before you start!

When paneling over masonry, such as concrete block, fix furring strips to the wall, using masonry nails.

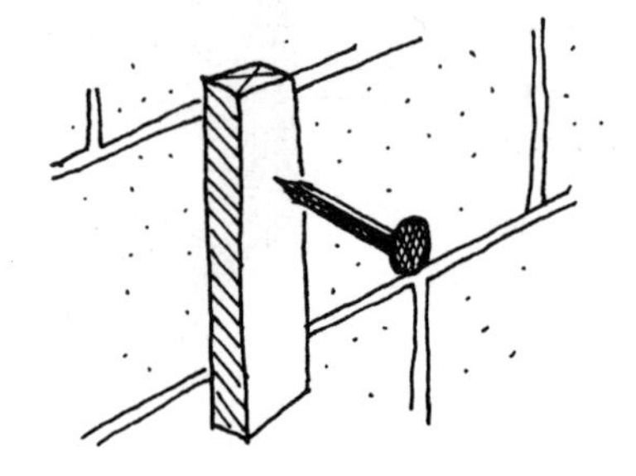

As with all other types of wall covering, do not forget to apply a vapor barrier first, such as a sheet of plastic.

When working with tongued and grooved material, blind nail through the tongues for best appearances, as shown below.

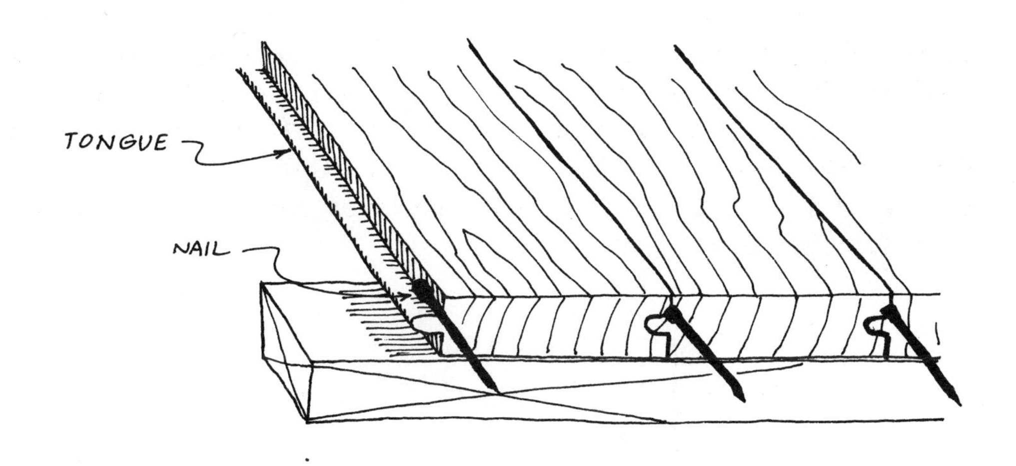

Do not try to hammer the nail all the way in; you will only mar the wood; use a nail set. Similarly, use a nail set when face nailing square-edged material.

Another way to disguise cracks caused by the wood shrinking as it dries out is to use V-grooved T & G (tongued and grooved) wood. Knotty pine is most common in this form; for closets and bathrooms, smaller T & G boards of cedar, which is very aromatic, are available.

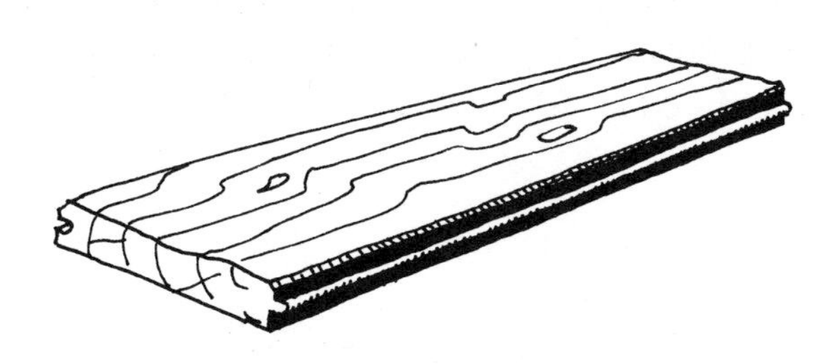

4. PLYWOOD PANELING

Plywood paneling, despite its imitative nature and relatively high cost per sheet, has become enormously popular with economy-minded builders, mainly because of the ease of application and consequent labor-saving.

Under the general heading of plywood paneling, I mean to include all kinds of plywoods, hardboards, and particleboards, all of which are available in a wide variety of thicknesses, textures, and patterns, but all of which usually are supplied in sheets measuring **4' × 8'** (**1.22 m × 2.43 m**), making application virtually the same regardless of type.

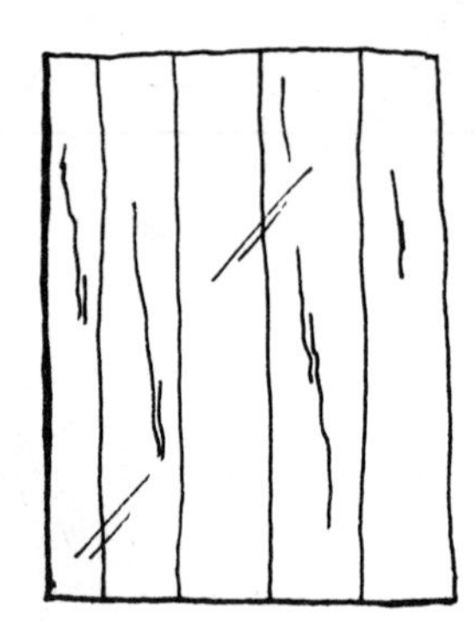

Order what you need delivered a week or so before using it and a week or so after any sheetrocking has been completed to allow the paneling to acclimatize to the house's moisture condition and the sheetrock to cure properly (this also produces a lot of moisture). Store the panels flat. Choose matching nails, preferably finish nails that can be set in the paneling grooves and blended in.

The most important point is to have adequate nailing surfaces before you start. The sheets, being **4' × 8' (1.22 m × 2.43 m)**, if applied vertically (as is usual), should have joints aligned with the studs, which are usually **16" (406 mm)** on center. However, because of doors and windows and other individual circumstances, the sheets will rarely line up with the studs. The solution is to install horizontal blocking every **4' (1.22 m)** between the studs.

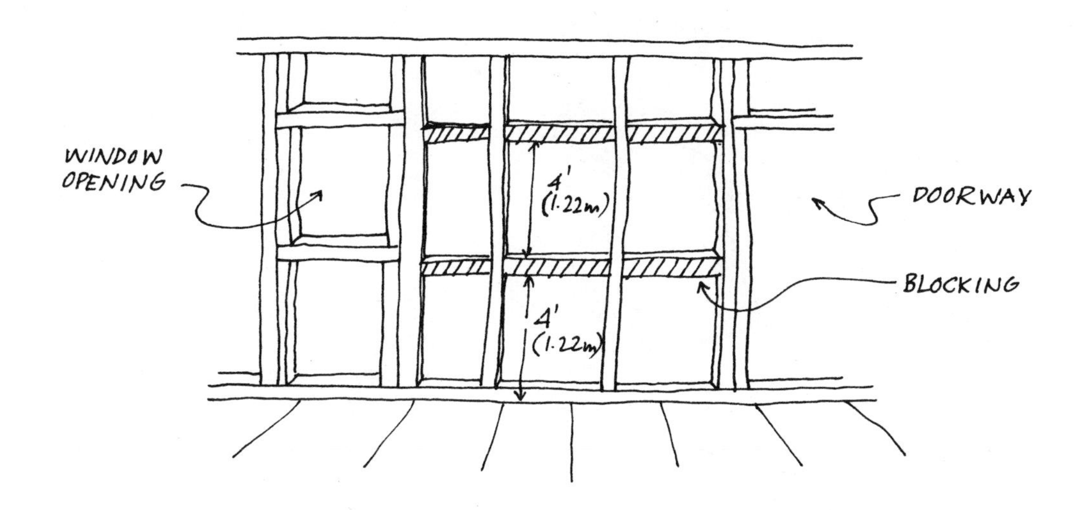

Unlike sheetrock, you can't hide chipped edges or corners, so be VERY careful. Try standing all the sheets up against the wall in the places they will go to check the resultant color matches and grain effects. Then start in one corner and work out.

Regardless of the wall, fix your panels perfectly plumb!

This will determine whether the room looks all right or not. Since most walls and ceilings do tend to vary a little (at least) from perfect plumb and perfect level, fix the first sheet with a small gap at the top (say **⅛" (3 mm)**) between it and the ceiling, and this will hopefully take care of any minor ceiling irregularities.

Most rooms are around 8' (2.43 m) high. When paneling rooms which are just a little higher, it is often easiest to panel down from the ceiling and hide the difference at the bottom with base board (called skirting in Britain).

When you come to electric outlets and switches, measure carefully - as with sheetrock. Drill a small hole to start the cut out, then use a keyhole saw or small electric jigsaw. A good method for getting the hole in the right place is to measure down from the ceiling and across from the next panel.

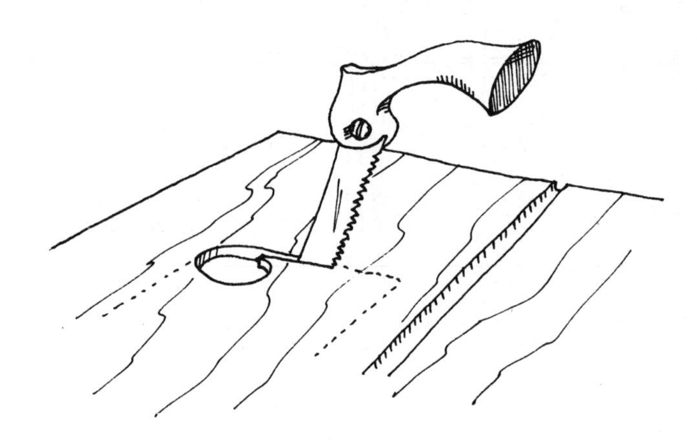

Avoid denting the panel when nailing. It is better to set the nail with a nail set. Nail at the top first - this way it is easier to adjust the sheet.

Try to arrange the panels so they join over an opening, as illustrated opposite; it is easier this way than having to cut the entire opening out of one panel.

WINDOW OPENING
WINDOW OPENING
THIS IS EASIER TO DEAL WITH THAN THIS

Leave all trim, such as ceiling moulding, door casing, etc., until all the paneling is up and all nails have been set.

Lastly, be very careful when moving the panels; watch all edges and corners all the time!

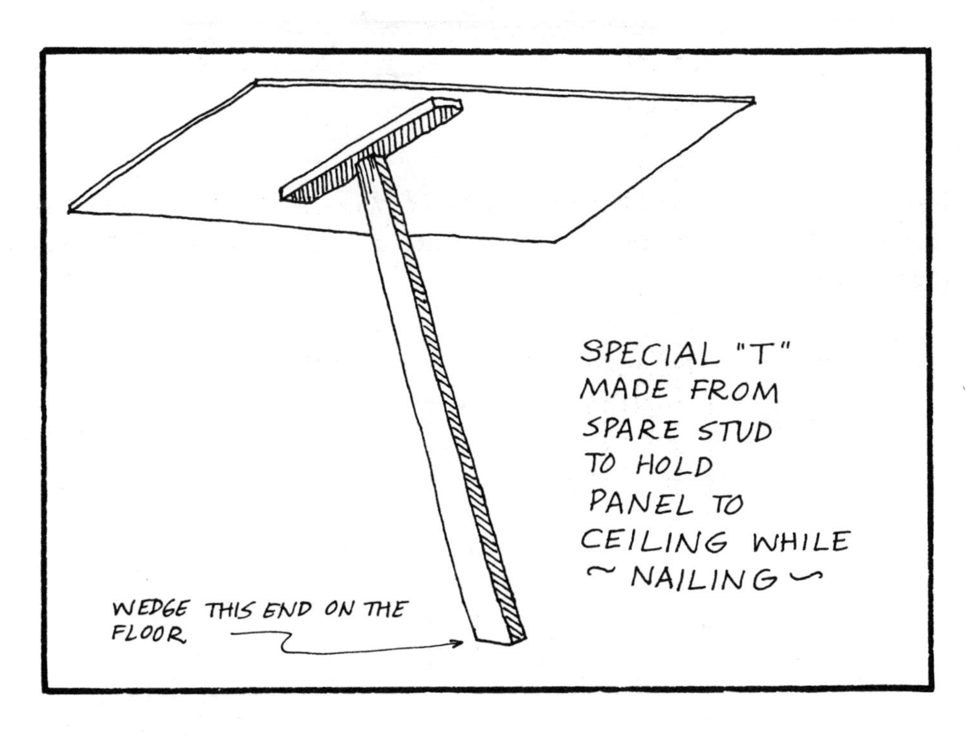
SPECIAL "T" MADE FROM SPARE STUD TO HOLD PANEL TO CEILING WHILE ~ NAILING ~
WEDGE THIS END ON THE FLOOR

5. TILED AND SUSPENDED CEILINGS

Ceiling tiles are manufactured in a variety of designs and standards, some mainly for easy application and others for sound-deadening and noise-reduction.

The easiest application is with the use of furring strips nailed to the rafters or joists of the floor above. For 12" × 12" (305 mm × 305 mm) ceiling tiles, simply nailing furring strips on 12" (305 mm) centers and either nail, glue or, best of all, staple the tiles to the furring strips

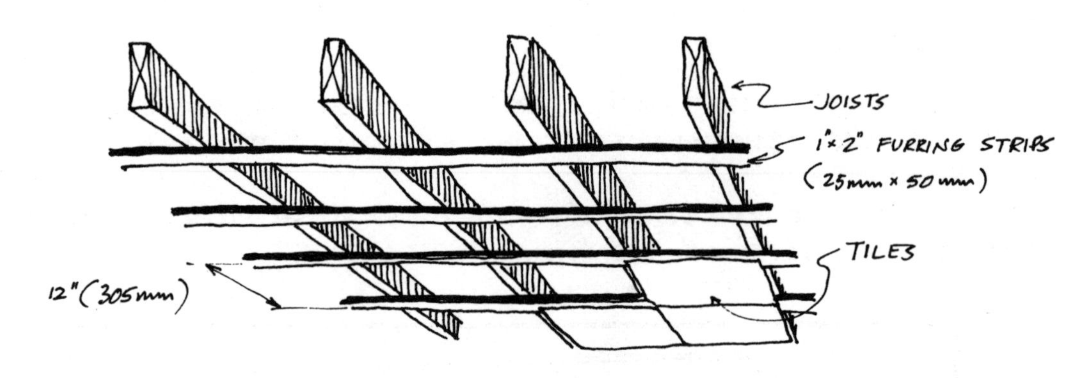

Arrange the furring strips so that the tiles lie symmetrically on the ceiling, having reduced tiles equal at both edges rather than at one side only.

YES

NO

The tiles may usually be cut with a utility or sheetrock knife. The edges of the ceiling may be supported by a ceiling moulding, so don't worry about furring strips where the ceiling meets the wall.

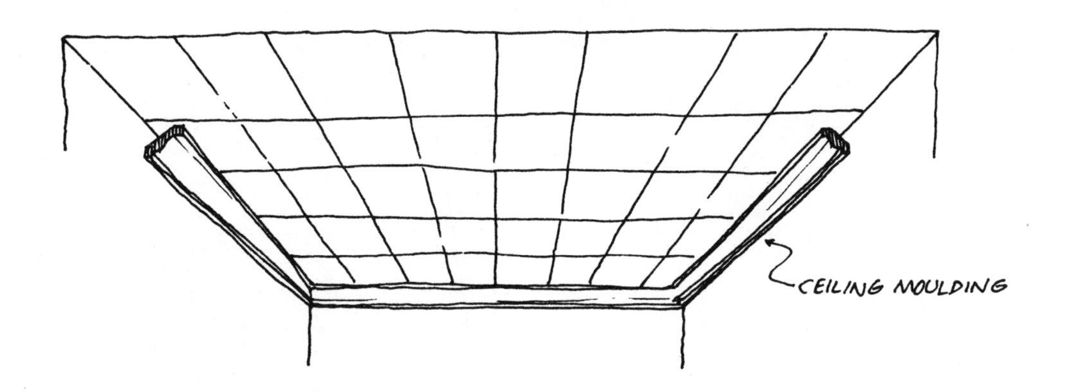

Suspended ceilings have the added advantage of being able to include flush lighting panels, making so-called luminous ceilings. Once again, there are numerous manufacturers, all with their own systems, but the basic procedure is similar in all cases.

Firstly, make an accurate plan of the ceiling to determine where the suspended runners which hold the panels (and sometimes the light fixtures – sometimes these may be fixed to the ceiling or joists directly above) will go. Normally these runners run parallel to the length of the room **2' (610 mm)** apart and at right angles to the ceiling joists.

Then plan the position of the cross-members, which are usually **2' (610 mm)** to **4' (1220 mm)** apart. With the basic grid thus sketched in, plan the lighting placement – if you are going to include lighting panels in the ceiling.

Now run whatever wiring is needed, install the lighting fixtures, if these are not to be suspended with the ceiling, and make sure that the ceiling provides nailing for the hangers which hold the runners by nailing blocks between the joists if necessary.

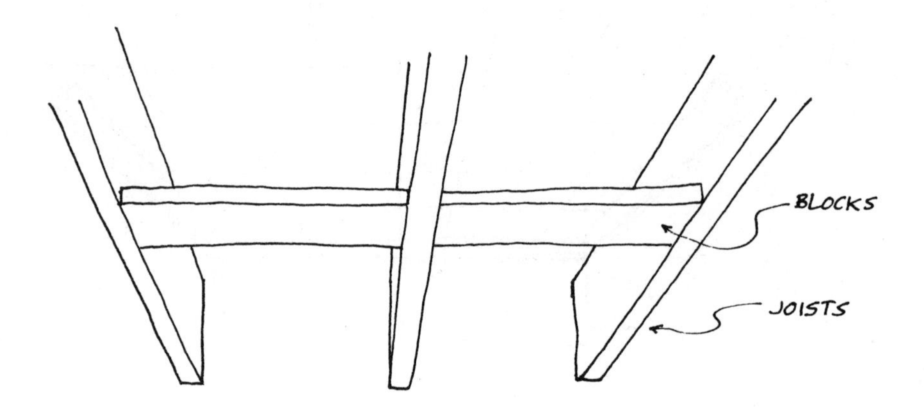

Next fix the wall angles so they are level and at the right height.

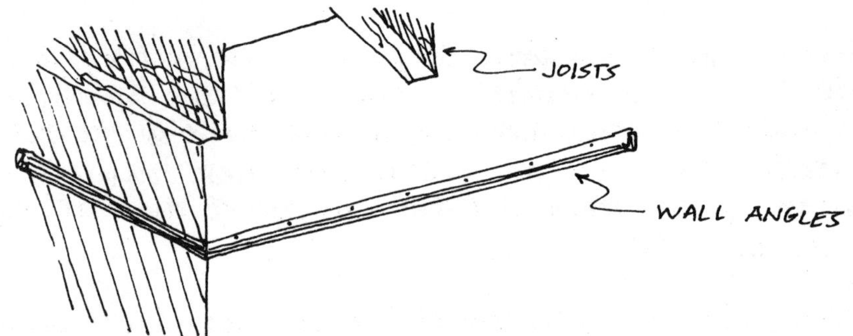

Then attach hanger wires to the joists (and blocks) and attach the main runners, after which the cross-tees are snapped in.

Finally, lay in the ceiling panels and lighting modules, and the system is complete.

The actual suspension units may vary from manufacturer to manufacturer, but basically they are all similar to the units shown below.

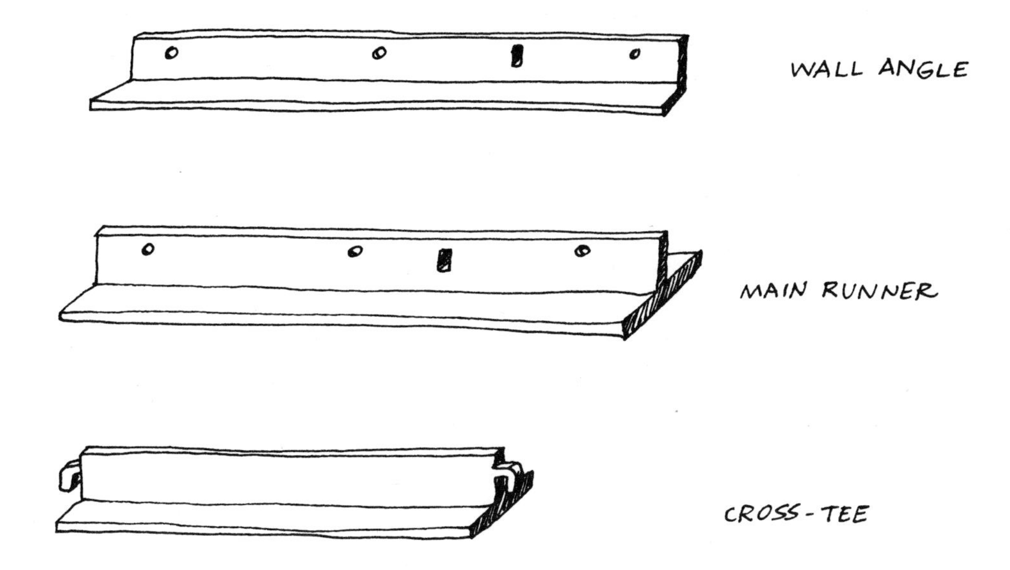

Chapter Four

Floors

There are numerous kinds of floors designed for various requirements, but they may be generally classified as either:

1. SINGLE LAYER FLOORS or
2. DOUBLE LAYER FLOORS,

and with respect to the form of joint, in the case of wood strip floors, as:

1. STRAIGHT OR BUTT or
2. TONGUE AND GROOVE.

Special feature floors, such as soundproof or fireproof floors, are not covered here, since we are dealing with a basically simple owner-built residence where "special features" are less likely to be required.

SINGLE FLOORS

In cheaper construction, a single finish board often suffices, especially where there is a cellar heated by a furnace to keep the cold and dampness out.

There are also other instances where a single layer floor is adequate, such as in the basement itself or in workrooms. In any case, with a single layer floor, it is imperative that all joints be over joists.

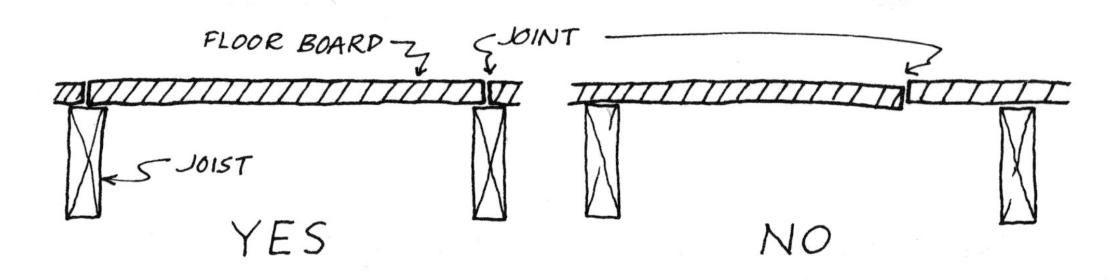

Carefully sorting the boards to length before starting to lay them can often save much waste and also allows control over possible unwanted contrasts in grain and color between adjoining boards.

Another form of single layer floor is a plywood floor subsequently covered with linoleum, tile, or carpet. This is both quick and cheap in terms of labor, but ultimately not as long-lasting or as versatile as a finish wood floor, although it doesn't matter if the carpet is replaced whenever it wears out.

DOUBLE FLOORS

Usually the floor consists of two layers:

1. THE ROUGH OR SUBFLOOR
2. THE FINISH FLOOR

Subfloors, which were formerly made of rough boards laid diagonally across the joists or of matched boards laid at right angles to the joists, are now most commonly made of plywood laid down in **4' × 8'** (**1.22 m × 2.43 m**) sheets.

The reason for laying the strip subfloor diagonally is to afford extra bracing, and because a finish floor is never laid in the same direction as a subfloor; but since the direction of a finish floor may change from room to room, diagonal subflooring is the only way to ensure a cross-laid pattern between subfloor and finish floor.

DIAGONAL SUBFLOORING

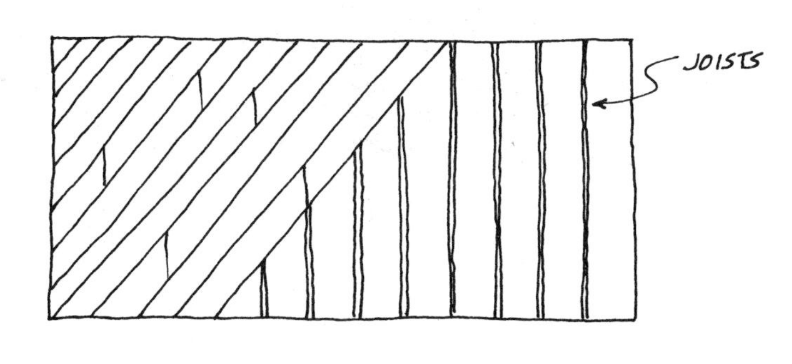

PLYWOOD SUBFLOOR

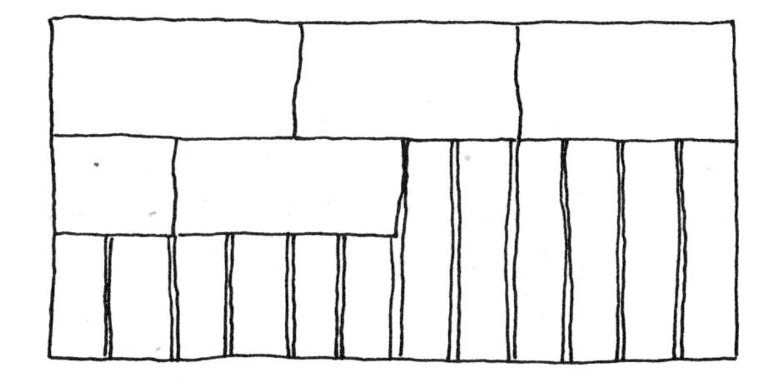

One constructional advantage of double floors is that the subfloor being laid early on, often before the framing of the walls is begun (though sometimes after), a floor for the builder to walk on is provided which will take rough usage, the finish floor, not being laid until after the painting and all other heavy work is finished, being thus protected.

In any event, the construction of subfloors is fully covered in my earlier book "**Illustrated Housebuilding**"; we are concerned here only with the finished aspects of flooring. Nevertheless, something which will pay dividends in securing a firm and squeak-free floor is to go carefully over the subfloor before laying the finish floor and make sure that it is flat, level, and well nailed. Squeaks occur because boards are not tight and consequently rub against one another. So if in doubt nail some more to make certain that every board is secure.

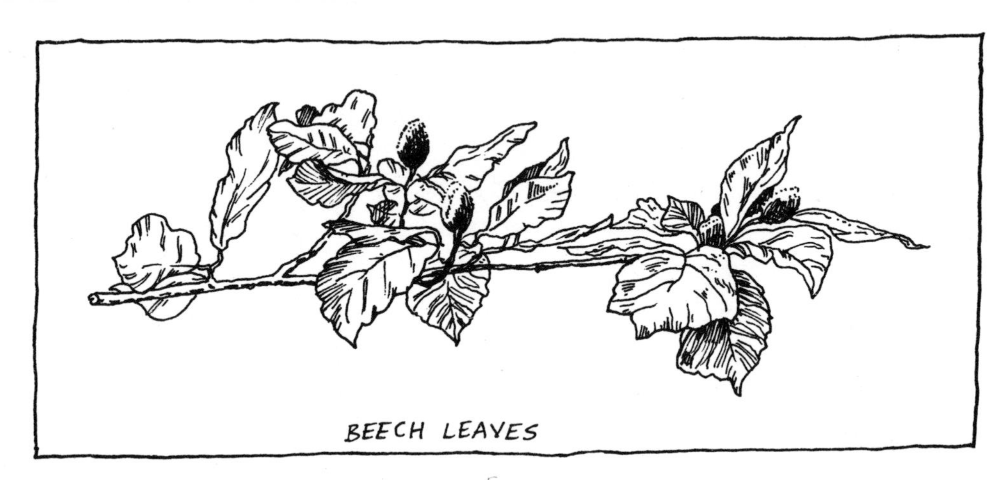

BEECH LEAVES

Between the subfloor and finish floor you should install a layer of **15** lb. building paper or deadening felt which stops dust and helps deaden the sound, and also, where the floor is over a basement or crawl space, will increase the warmth of the floor by preventing air infiltration.

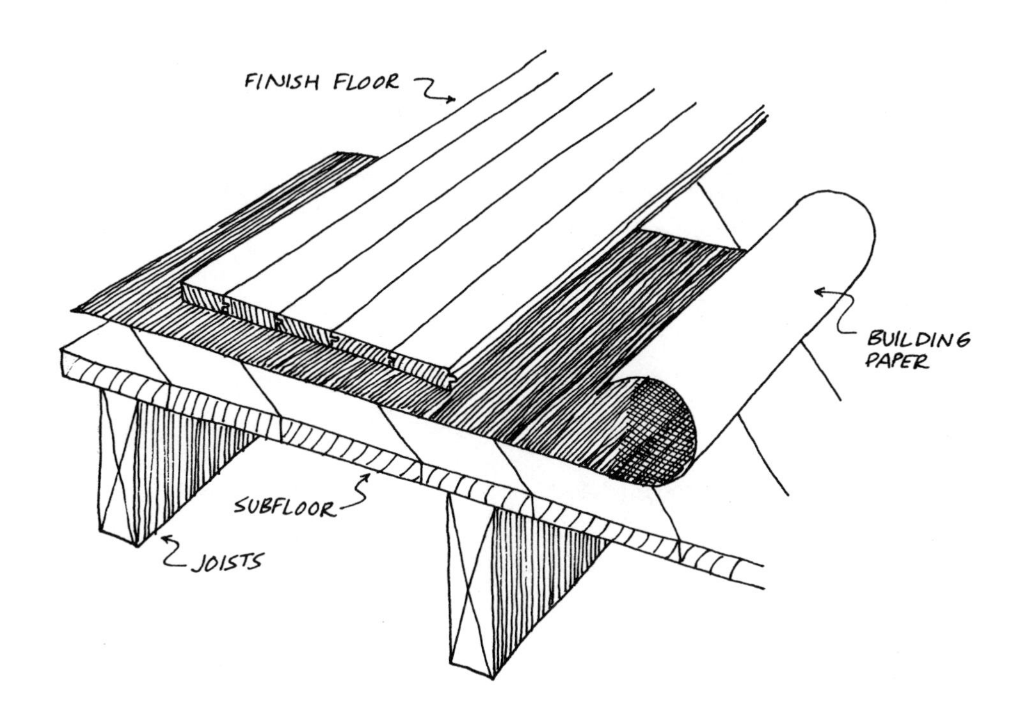

When you have laid the paper (which is black) it will be found very helpful to chalk in the location of the joists so you know where to nail the finish floor.

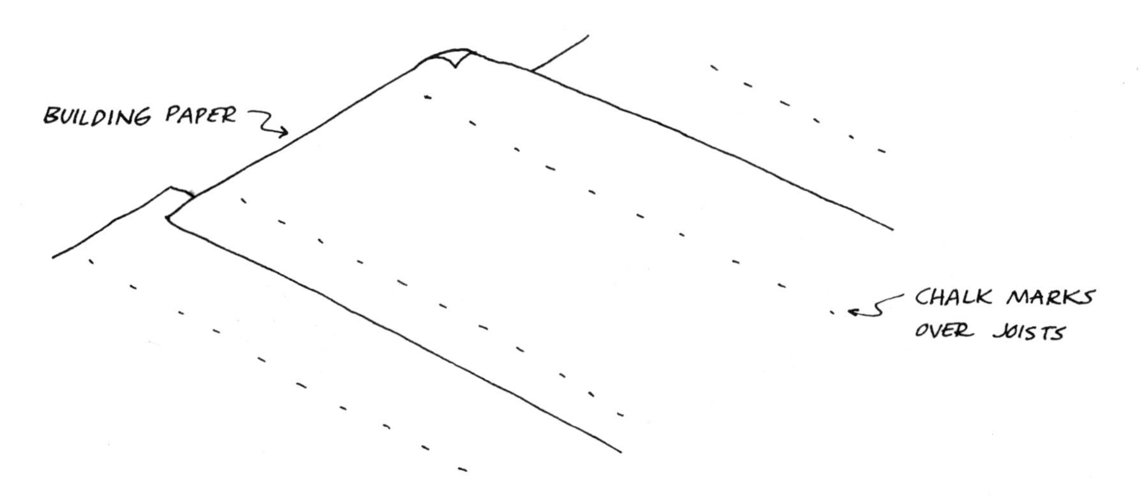

The finish floor may consist of hardwoods or softwoods laid in various ways.

Softwood floors and strip flooring in general are usually laid crosswise to the joists and lengthwise to the room for best appearances. The softwoods now most commonly used are Douglas fir, redwood, western larch, and hemlock and white pine.

The commonest hardwoods are oak and maple, although where expense is not a limiting factor birch, yellow pine, walnut, red gum, cherry, or mahogany can also be used (as may be many other exotic and expensive hardwoods).

Hardwoods, when used in strips or boards, are similarly usually laid crosswise to the joists, but when used in short lengths, may be laid in various patterns, as in parquet.

RANDOM WIDTH FLOOR OF PINE BOARDS

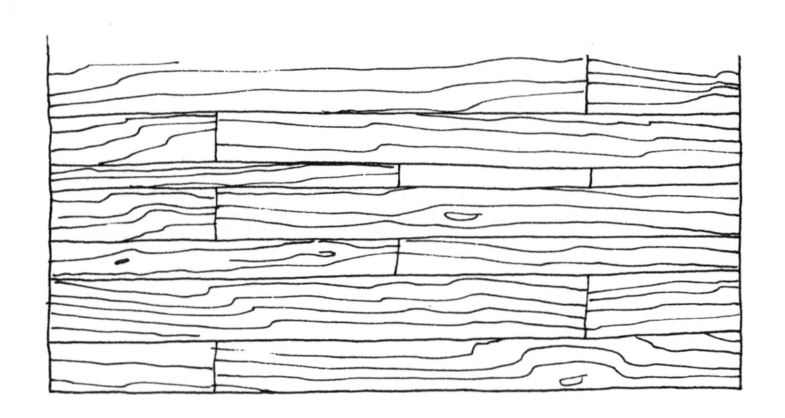

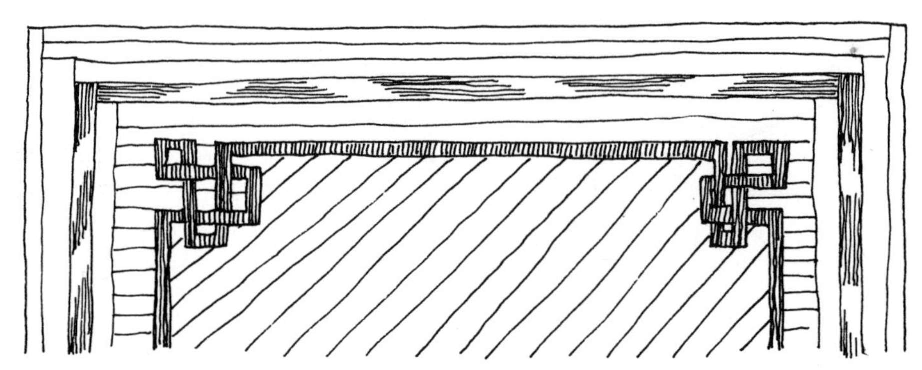

PATTERNED FLOOR OF OAK & CHERRY (BARLBY CASTLE, KENT)

Before the twentieth century, square-edged single floor boards were the rule. Thickness varied, but was generally greater than is used today. Also, the widths were usually much greater too, sometimes up to as much as 36" (915 mm).

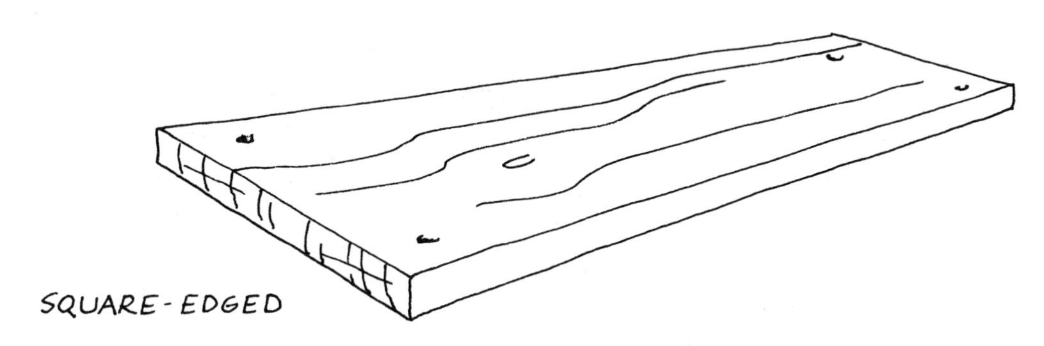

The boards were face-nailed or pegged with hardwood pegs. Later the boards were rabbeted, and gradually tongued and grooved boards became the rule.

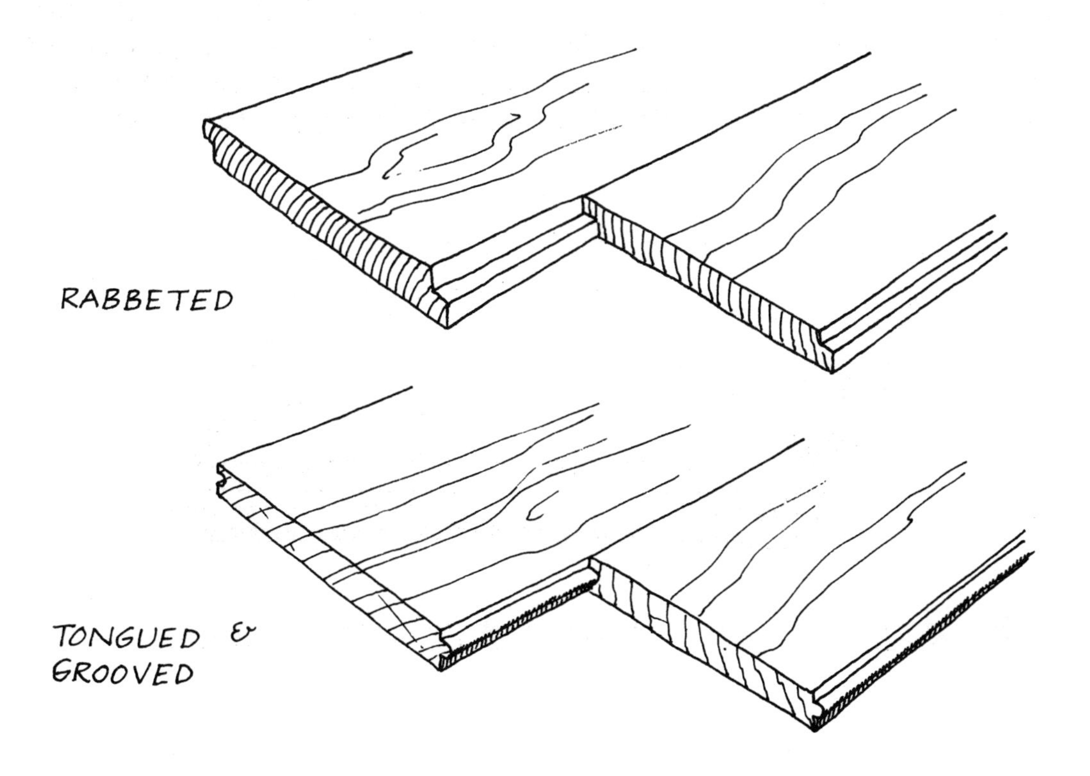

Nowadays, when wide boards are used, screws are often set deep in the wood and the holes filled with circular wooden plugs.

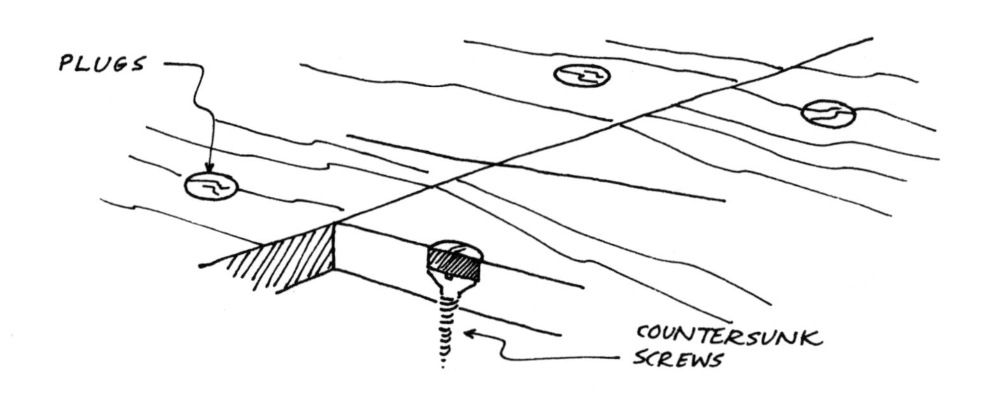

The very best floor would be made of seasoned quarter-sawn hardwood, end-matched (this means every board is tongued and grooved on the ends as well as the sides), and hollow-sawn on the back to ensure firmer contact with the subfloor, since there is less surface which can be affected by any subfloor unevenness.

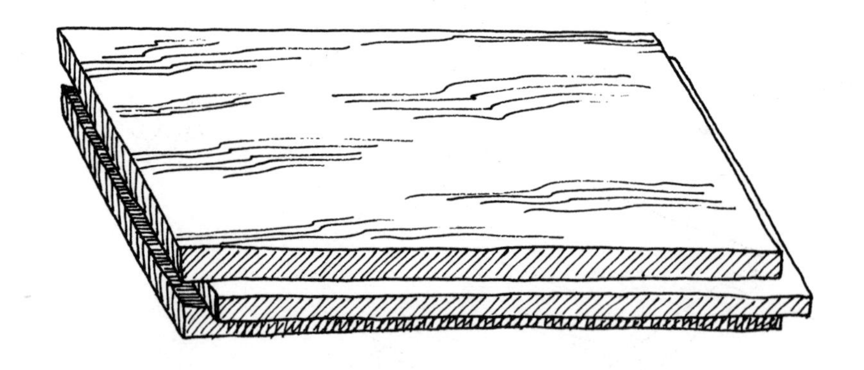

QUARTER-SAWN END-MATCHED OAK

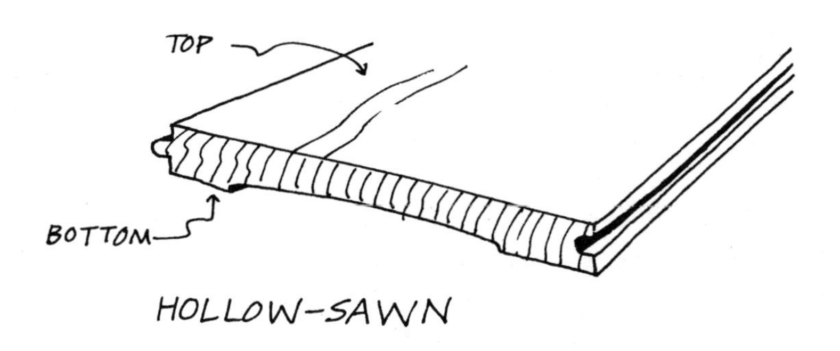

HOLLOW-SAWN

It should be borne in mind, however, that with regard to how the board is sawn from the tree, plain-sawn, being cheaper because it is an easier cut, is more common than quarter-sawn.

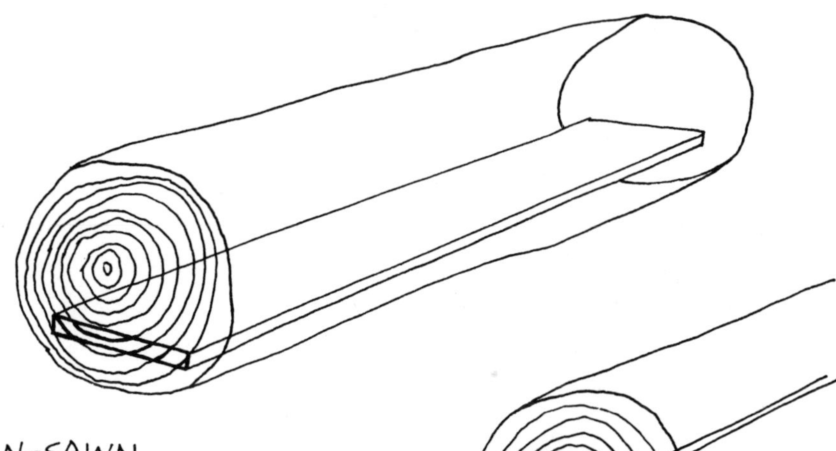

PLAIN-SAWN

The boards are all sliced from the tree in one plane; consequently, the grain runs across the width of the board, making it easier to warp.

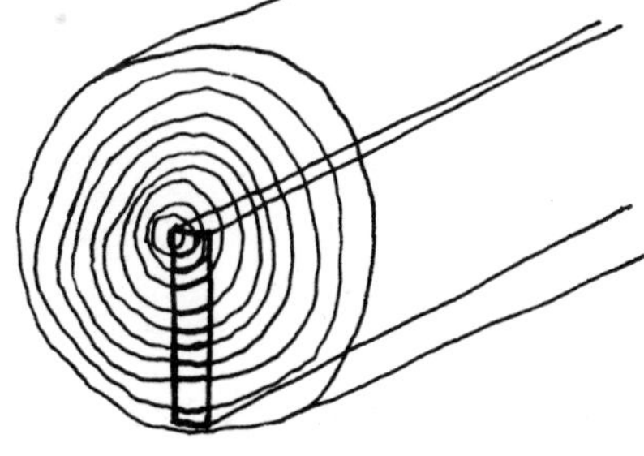

QUARTER-SAWN

All the boards are sawn from the center to the outside; thus the grain always runs at right angles, making the board more stable in shrinkage.

Softwood flooring is usually not end-matched and is usually wider - up to 8" or 10" (203 mm or 254 mm) is common - whereas hardwood flooring is more usual in widths from 1½" to 4" (38 mm to 107 mm), and also in shorter lengths.

The best way to obtain softwoods is to have a sawmill tongue and groove native wood, and mill one side only. This gives you a thicker board than one milled on both sides and is a lot cheaper. However, you will have to season it yourself; whereas hardwood flooring may be obtained in prefinished bundles which are supposed to be kiln-dried to the correct moisture content.

In any event, finish flooring should never be laid in a house which is damp, or which has been unheated during the winter months. Kiln-dried wood will quickly absorb moisture and swell, only to shrink again when dried out, leaving gaps between the joints.

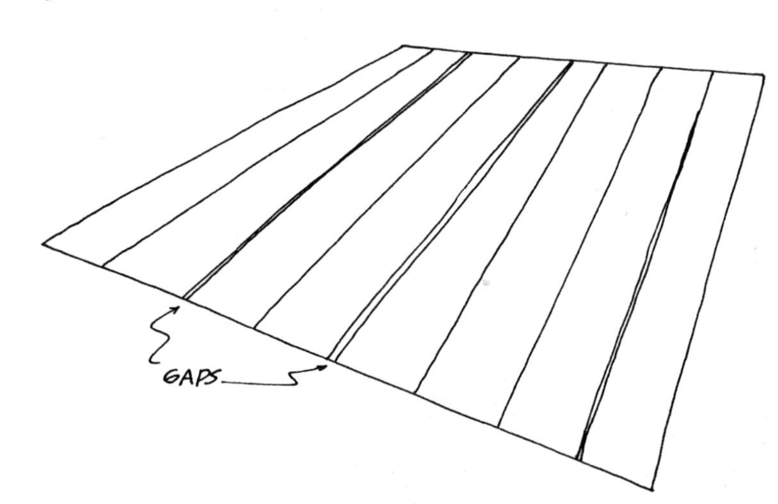

GAPS CAUSED BY SHRINKAGE CAUSED BY FLOOR HAVING BEEN LAID WITH A TOO HIGH MOISTURE CONTENT

Storing the flooring in the house as long as possible before laying it (assuming the house is heated, if it is winter) is ideal. If you are laying softwood lumber obtained directly from the mill rather than the lumberyard, you should try and keep it at least six months before laying it.

LAYING A FINISH FLOOR

The first problem, having decided what kind of floor you are going to lay, is how much to order, because **100 sq. ft. (9.29 sq. m.)** of flooring does not cover **100 sq. ft. (9.29 sq. m.)** of floor area! This is because of the difference in actual and nominal measurements and because a certain amount of wood is invariably lost at joints and ends.

The smaller the units you are laying, the more you should add. For example, if you plan to floor a particular room, discover the area by multiplying the length of it by the breadth of it and add **25%** if the flooring is to consist of **2" (50 mm)** wide boards, or **33⅓%** if it is to consist of **1½" (38 mm)** boards. This will give you enough if the room is rectangular – if there are bay windows, closets, or other nooks, add proportionately more.

2" (50 mm) BOARDS:
LENGTH × BREADTH + 25%

1½" (38 mm) BOARDS:
LENGTH × BREADTH + 33⅓%

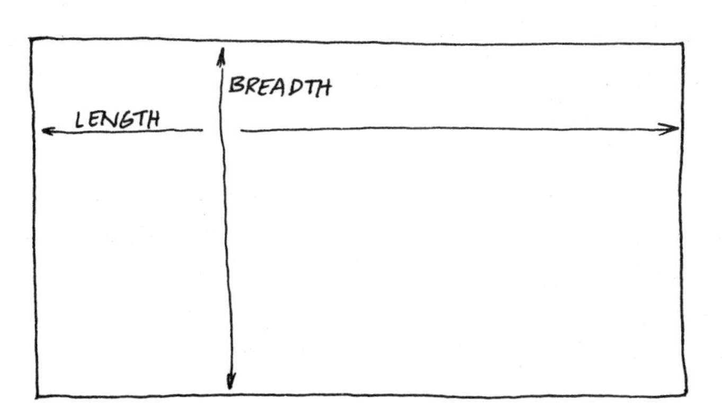

COMPUTING AMOUNT OF FLOORING NEEDED

As stated before, make sure the subfloor is level and tight. Do some extra nailing if necessary and see that all nails are well set. Lay down a good quality building paper.

Choose the right nails for the floor you are laying. For example, 25/32" (20 mm) thick matched tongue and groove flooring is best blind-nailed with 8d. wire nails or cut flooring nails.

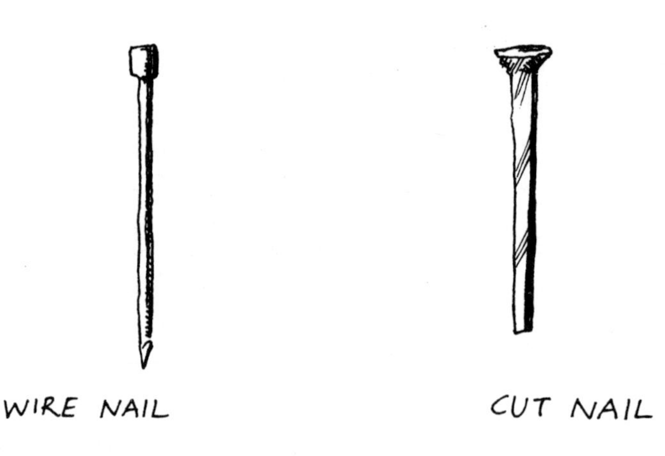

WIRE NAIL CUT NAIL

For ½" (12 mm) thick flooring, 6d. nails may be used.

Cut nails are better for use with hardwoods, since they punch rather than spread the fibers and are thus less likely to split the wood. Pre drill whenever you think there is any danger of splitting - it only takes a few moments to drill, and even though it may upset your rhythm, the split board could lie there for years!

The floor should be begun and finished ½" (12 mm) from the wall to allow for any expansion and thus prevent any buckling. This gap is hidden either by the baseboard or the baseboard shoe as shown opposite. Therefore the baseboard should ideally be removed before you start.

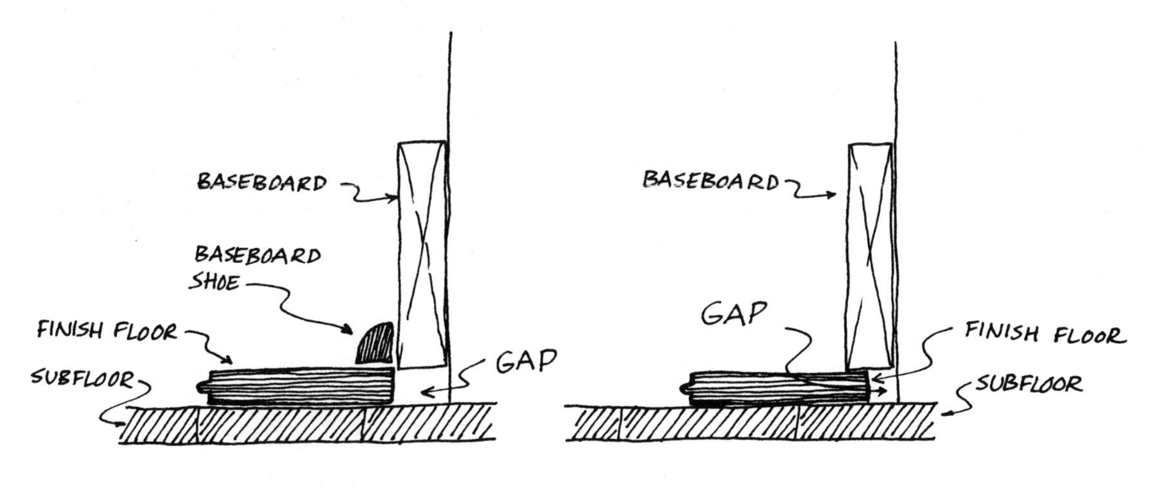

LEAVING A GAP AND CONCEALING IT

Also, before you start, remove any doorway thresholds.

The first strip, or board, is laid with the grooved edge facing the wall, and is face-nailed along that edge and blind-nailed through the grooved edge.

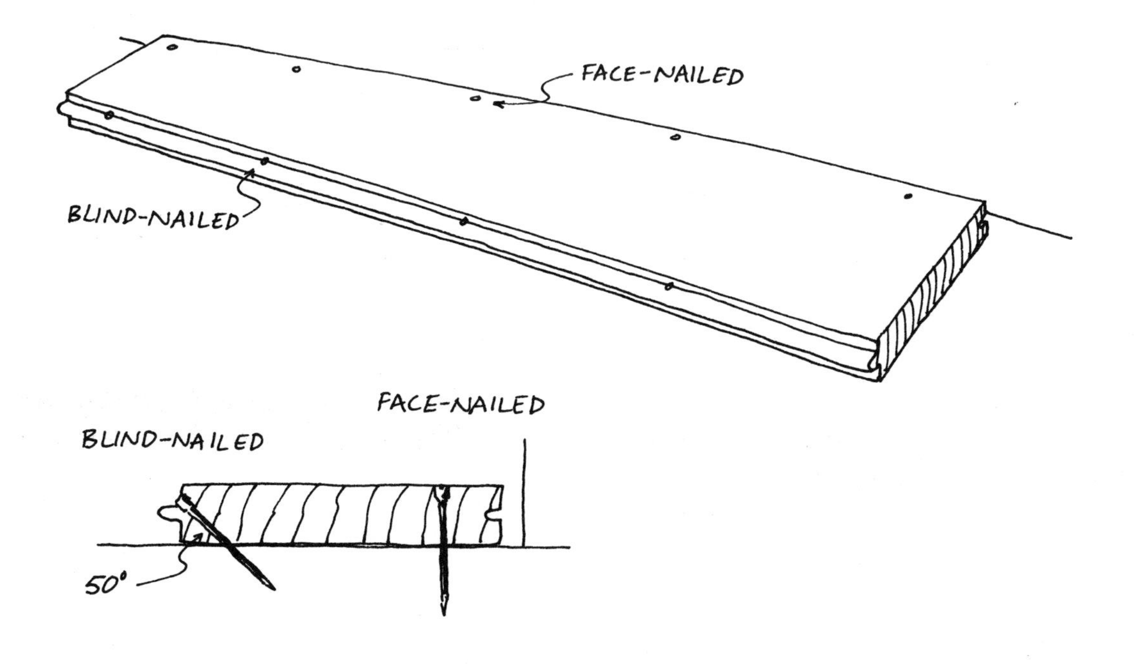

The tongue of one strip fits into the groove of the next strip, snugly but not too tightly or the floor could buckle under damp circumstances.

Make sure that the first strip is straight by sighting along it and from time to time, as you approach the opposite wall, measure the width of floor laid for variation.

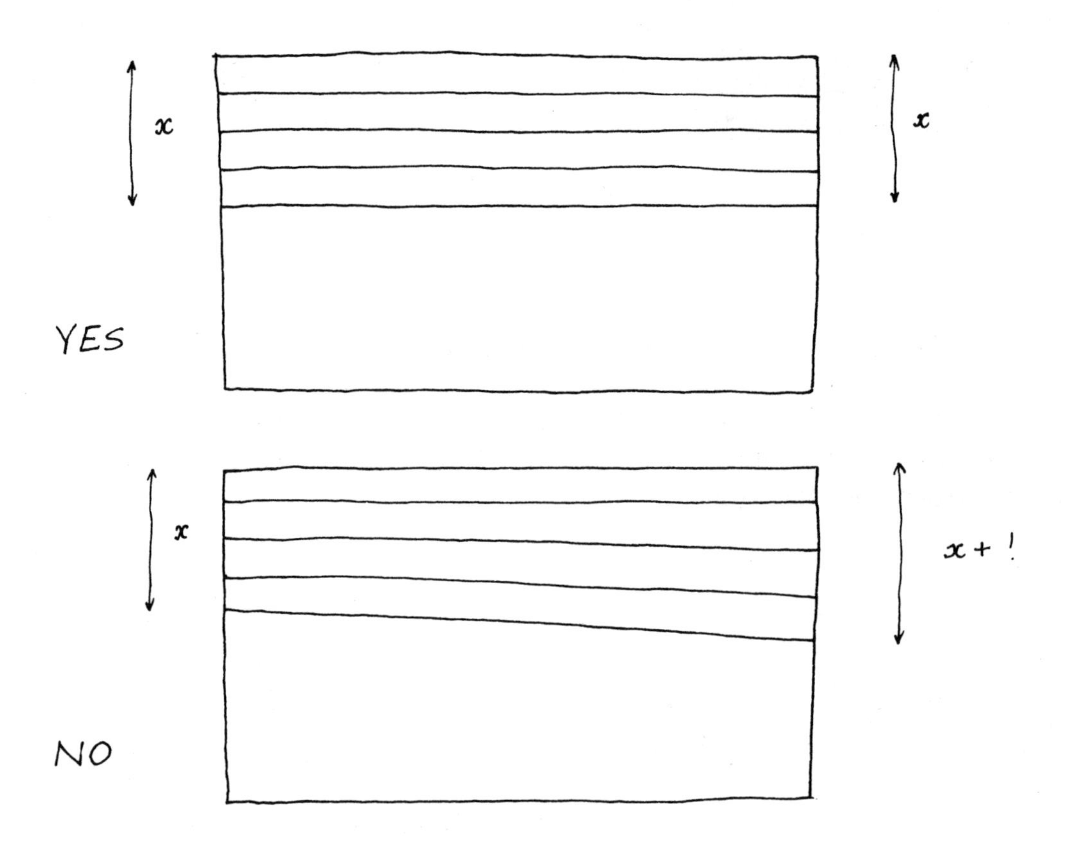

Nail to every joist, and when blind-nailing, do not strike the wood with the hammer or attempt to set the nail with the hammer alone; use a nail set.

Narrow width strip flooring may be laid three or four strips at a time and then driven tightly against one another.

Another method is to use a hammergun, which fits against the tongue of the laid board, and, being struck with a hard rubber mallet, drives a nail in at the right place and at the right angle. These guns may usually be rented from lumberyards where flooring is sold.

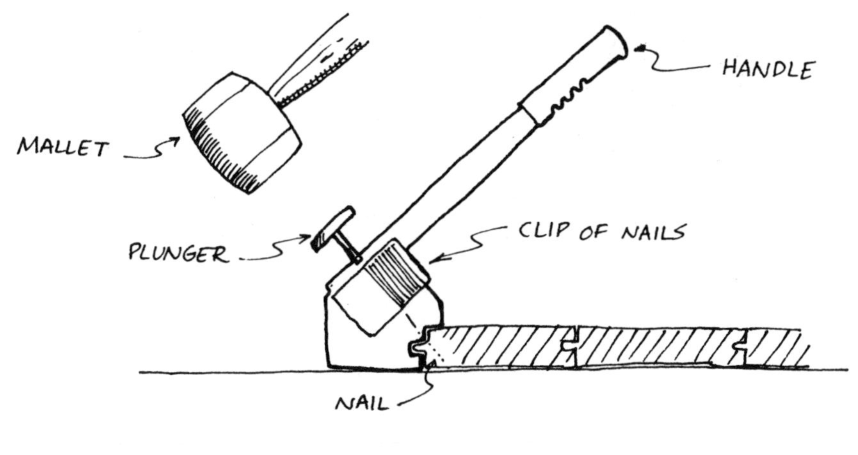

NAIL GUN

Since hardwood flooring is usually supplied in random lengths, mix the long with the short over the whole area - unless the center of the floor is to be covered by a rug, in which case, save the long pieces for the outside where they will show. It is easier to use long pieces for doorways and openings, while short pieces may be used to advantage in closets.

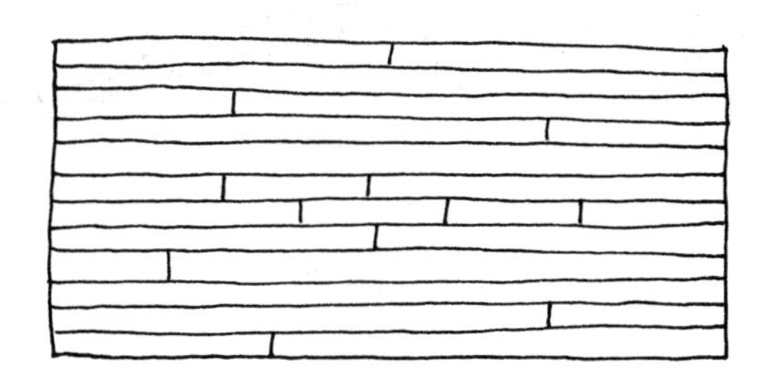

To drive the pieces tightly together, use a scrap piece of flooring and hit this rather than destroy the tongue of the strip being laid.

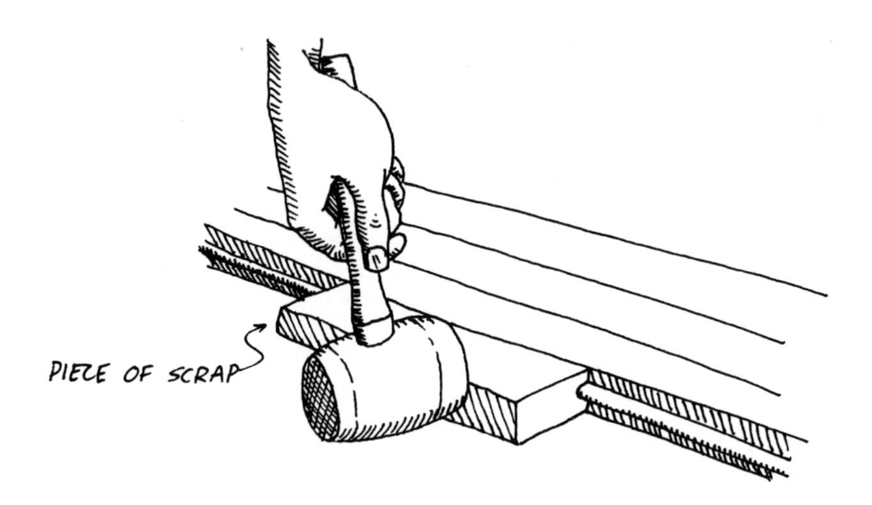

With wider boards, or longer bent pieces, the method is to nail a piece of scrap to the floor a little way ahead of the finish floor and drive a wedge in between, thus forcing the boards tightly together.

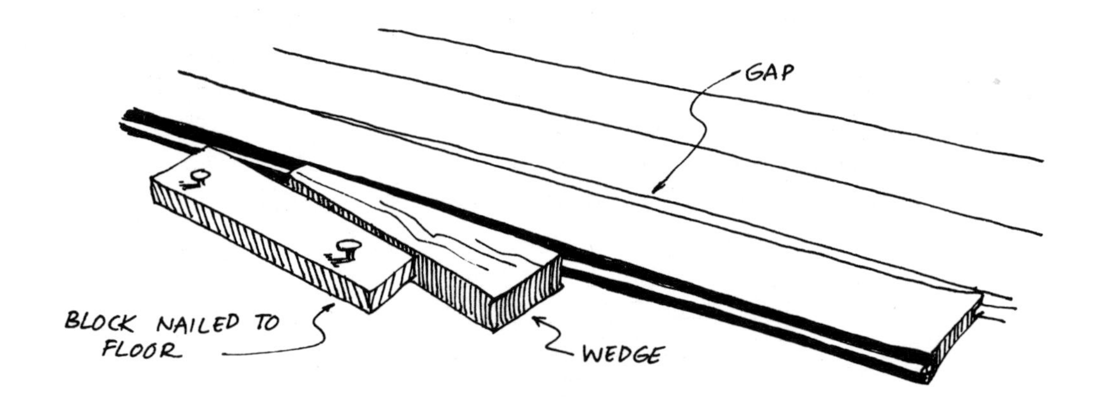

HOW TO CLOSE GAP

All butt joints (the ends of the boards) should be staggered. This is equally important for both wide and narrow boards.

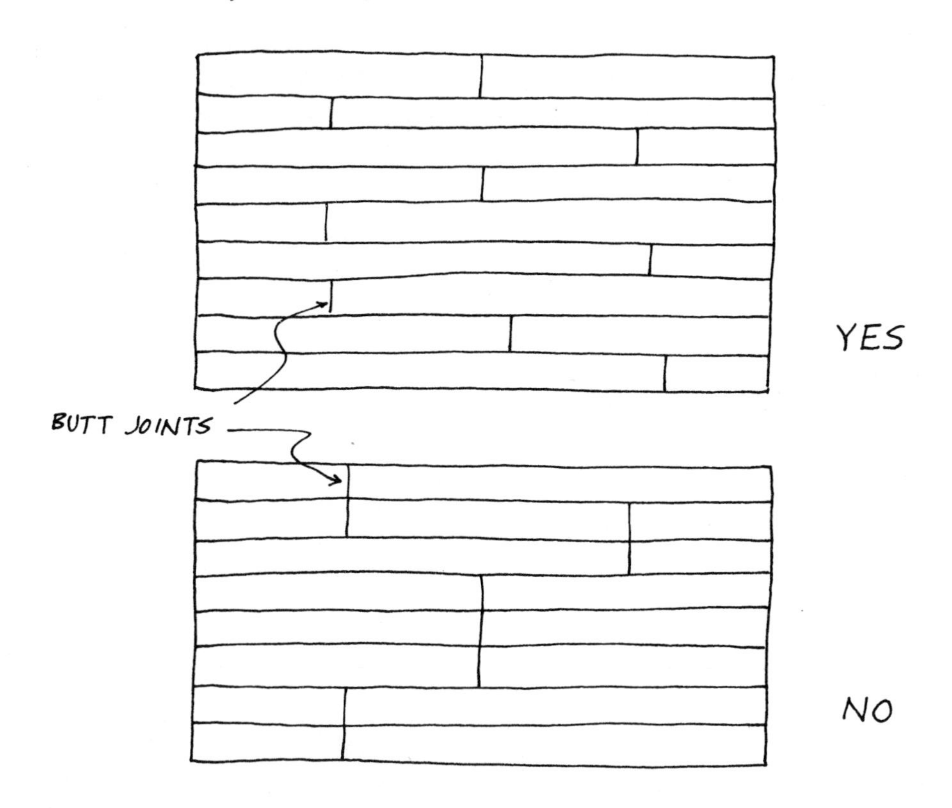

To lay boards in tight corners, and to lay the last few boards, fit them together as shown and press them into place by standing on them.

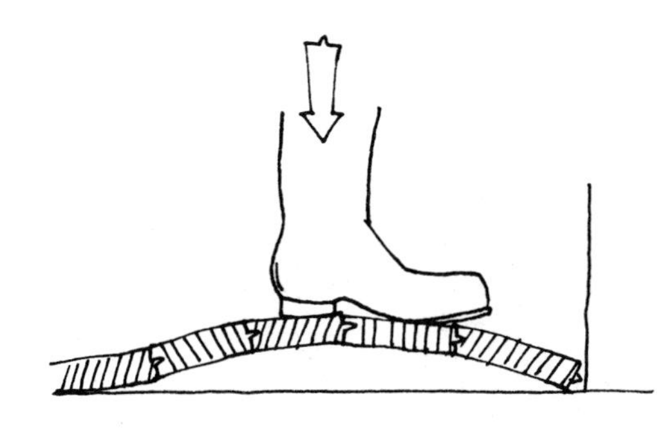

FITTING THE LAST FEW BOARDS

When all the floor is laid, replace all the trim, thresholds, and baseboards you may have removed.

When replacing the baseboard shoe, nail the shoe to the floor, not to the baseboard, or the baseboard, if it should shrink, might pull the shoe up from the floor and leave an unsightly gap.

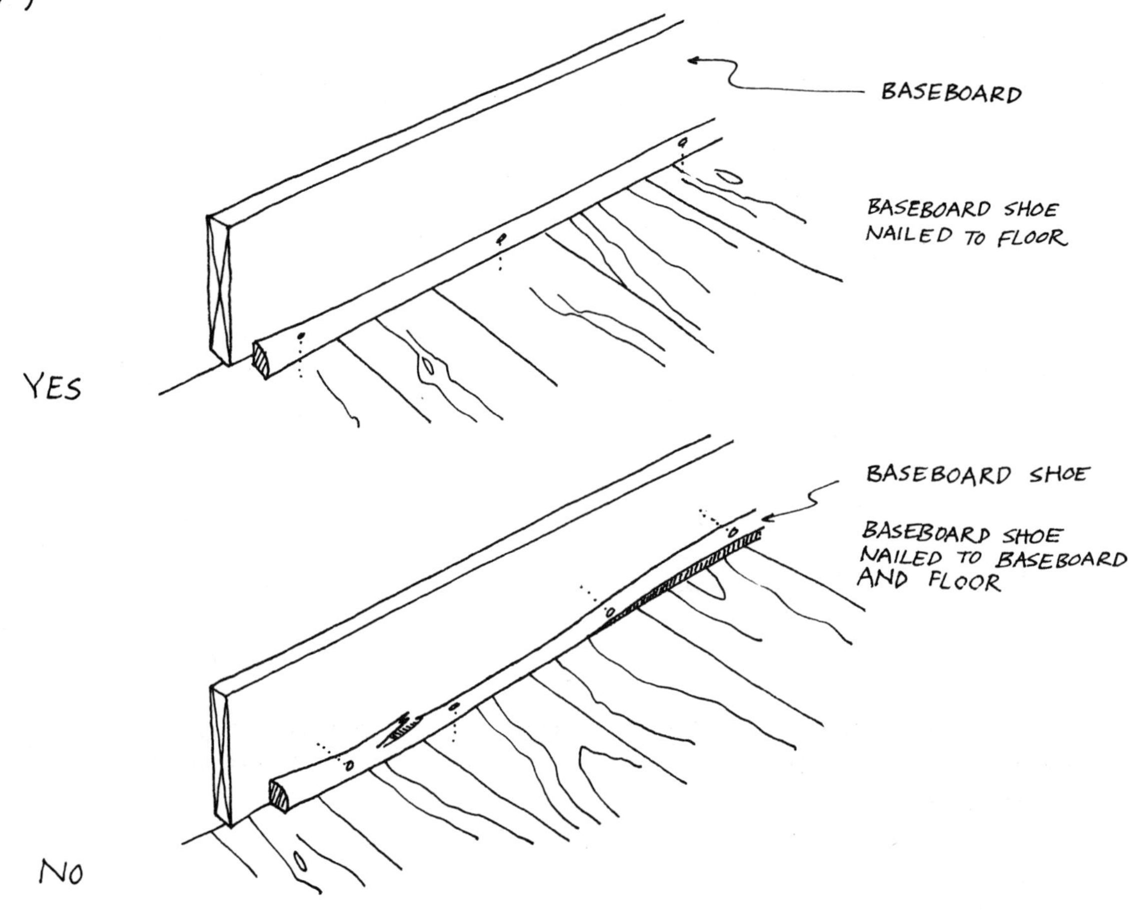

The shoe is most conveniently made from "quarter round." The actual baseboard may be anything from mill-made moulding or plain 1" × 2" (25 mm × 50 mm) to an elegantly moulded skirting board up to 12" (304 mm) high.

In all cases nail the baseboard to the studs. Cope the inside joints and miter the outside joints.

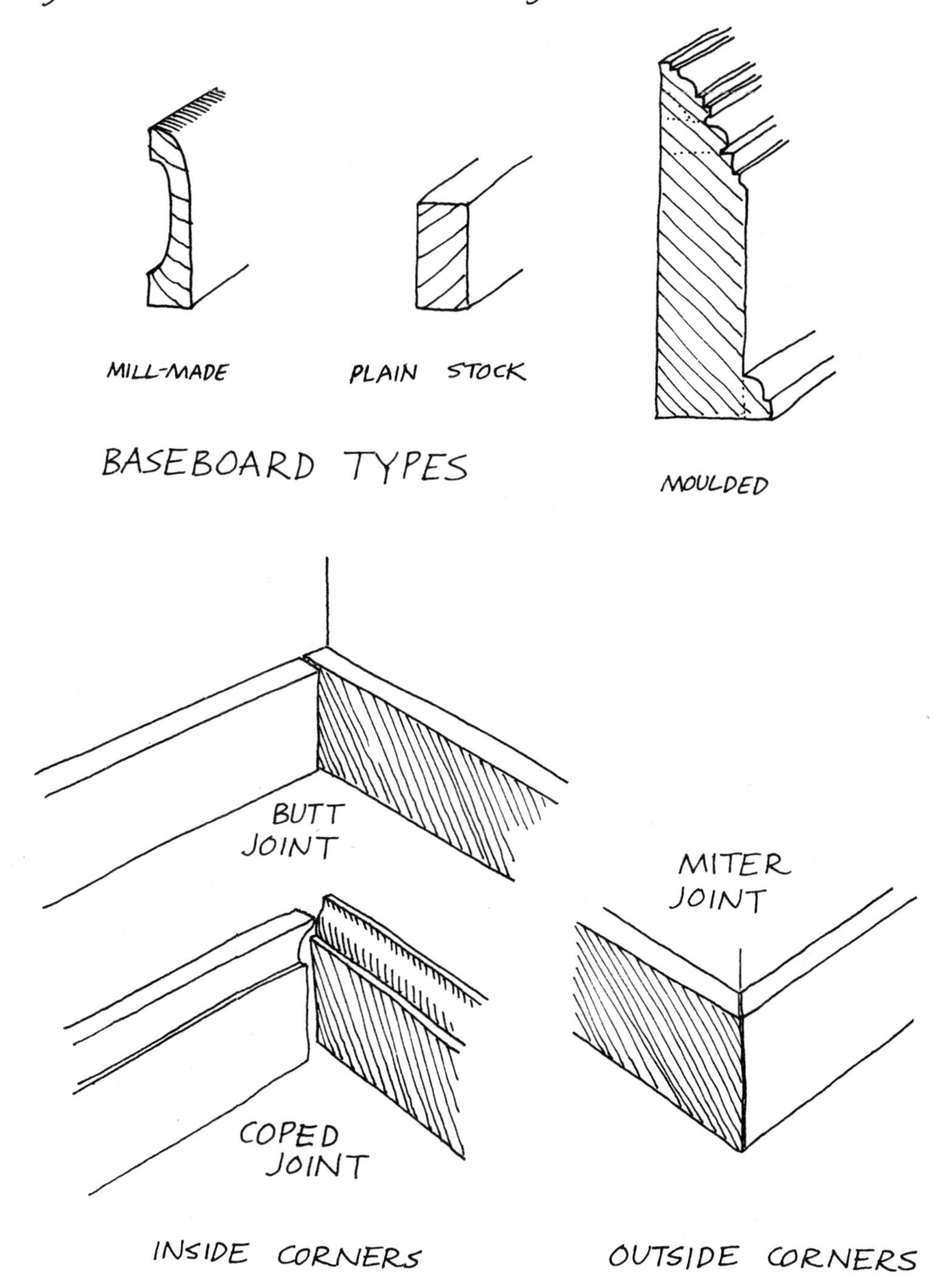

Chapter Five

Stairs

The construction of stairs is, like the framing of roofs, one of the most difficult aspects of housebuilding, calling for the most amount of knowledge and the greatest degree of skill.

It is virtually an art in itself, since hardly any two staircases are exactly alike, each usually being built to accommodate a unique set of circumstances, each staircase thus presenting its own peculiar problems.

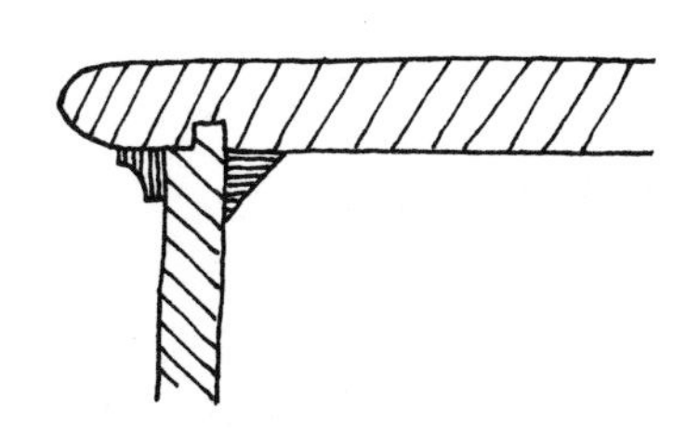

A whole volume could be written detailing the intricacies of stairbuilding, and yet the aspiring stair builder would still need much practical experience before he could call himself properly competent.

This section of this book does not therefore claim to be exhaustive, merely to present enough information so that the general principles may be grasped and thereby enable the builder to construct a straightforward staircase for a small residence.

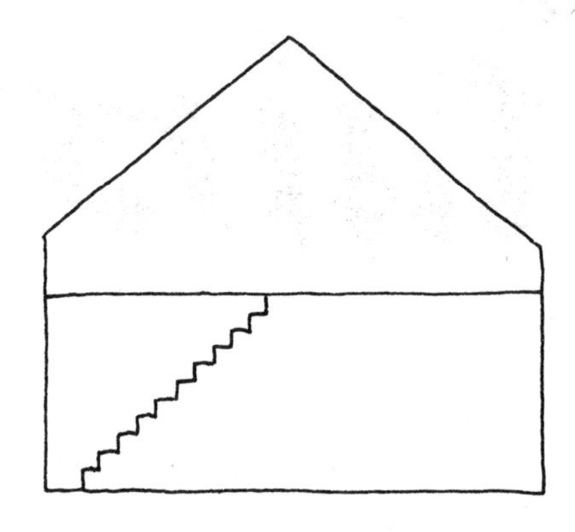

If your situation demands a more complicated solution than is here provided and you do not intend to become a professional stair builder, then it would be better to contract the work and concentrate your energies elsewhere.

Nevertheless, there are sufficient opportunities for relatively simple stair construction, such as stairs for porches, attics, basements, and sometimes even main stairs where the design is not too complicated, to warrant some discussion of stair building.

Provided the theory and principles are thoroughly understood and careful workmanship is practiced, this work should not be beyond the builder or carpenter who has managed to construct the rest of the house.

More than any other aspect of housebuilding, stairbuilding has its own large vocabulary of terms which must be understood in order that the subsequent instructions may be followed. It is important therefore to familiarize yourself with the following definitions before proceeding further.

DEFINITIONS

STAIRS are a series of steps which afford communication between different levels.

STAIRCASE is the term properly applied to the space or enclosure occupied by the stairs and landings; although it is often used synonymously with "stairs."

A **STEP** is one of the footspaces or platforms in a flight of stairs. In wooden stairs the step usually consists of a **RISER** and a **TREAD**.

TREAD
RISER
TREAD
RISER

TWO STEPS WITH TREADS & RISERS

Sometimes the riser is omitted.

FLYERS are steps in a flight of stairs that have rectangular treads and which are consequently parallel to each other as compared to WINDERS or TURN-STEPS, which are used in turning around a newel and which in consequence may have irregular or triangular treads, and which are not parallel to each other.

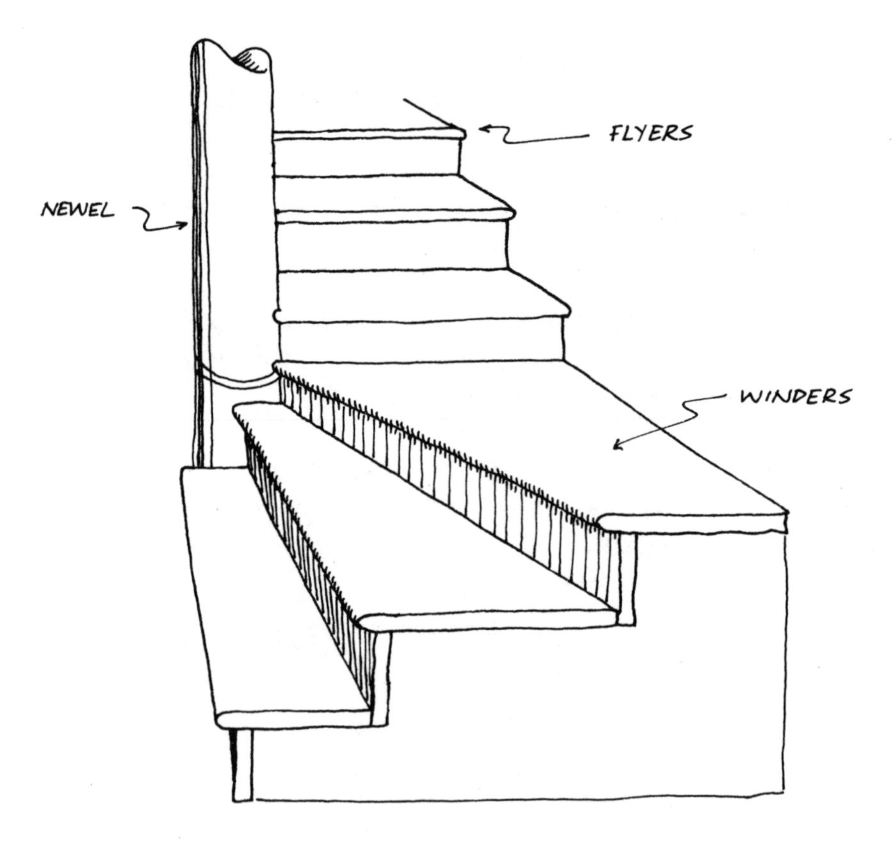

The outer end of the lowest step is often rounded, and is then known as a ROUND-ENDED or BULL-NOSE STEP; if the end is formed into a part of a spiral, it is known as a CURTAIL STEP.

OPEN STRING STAIR WITH BULL-NOSE STEPS

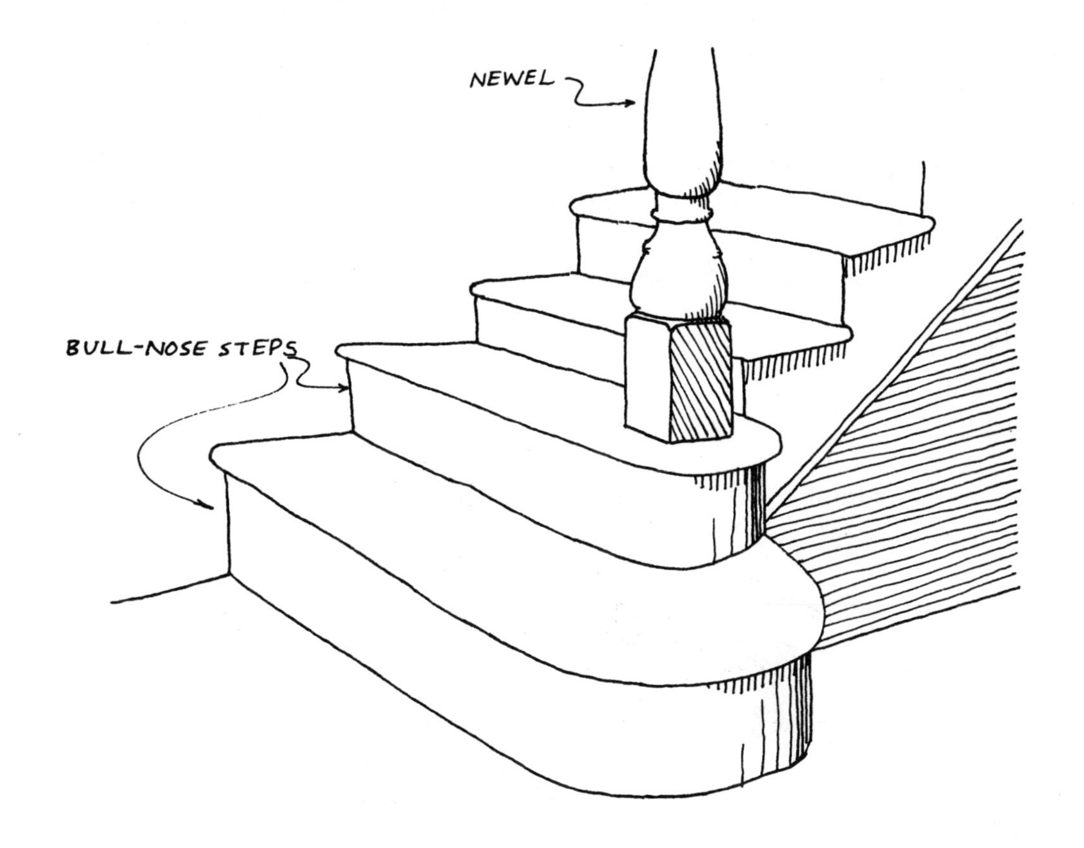

The NOSING or front edge of a tread may be SQUARE or ROUNDED or otherwise MOULDED. A straight edge laid on a flight of stairs to touch the nosings will give the LINE OF NOSINGS.

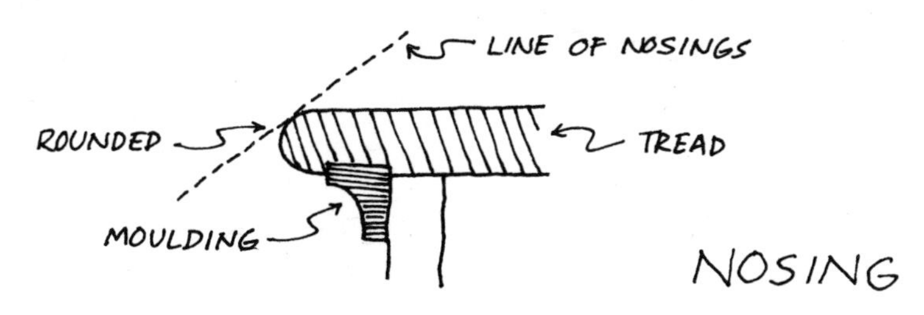

NOSING

STRING BOARDS, **STRINGS**, or **STRINGERS**, as they are now most usually called, are the sloping boards which hold the ends of the treads and the risers. In the illustration below, "**a**" is a section through a series of steps and shows one of the stringers in elevation; the treads and risers are housed in grooves cut in the stringer and held secure by wedges which are driven further in whenever the stairs begin to creak (because of the wood shrinking as it dries).

The inside angle formed by the tread and the riser is known as the **SOFFIT ANGLE**, and is supported by glued wedges as shown.

At "**b**" the same stringer is shown with the steps omitted.

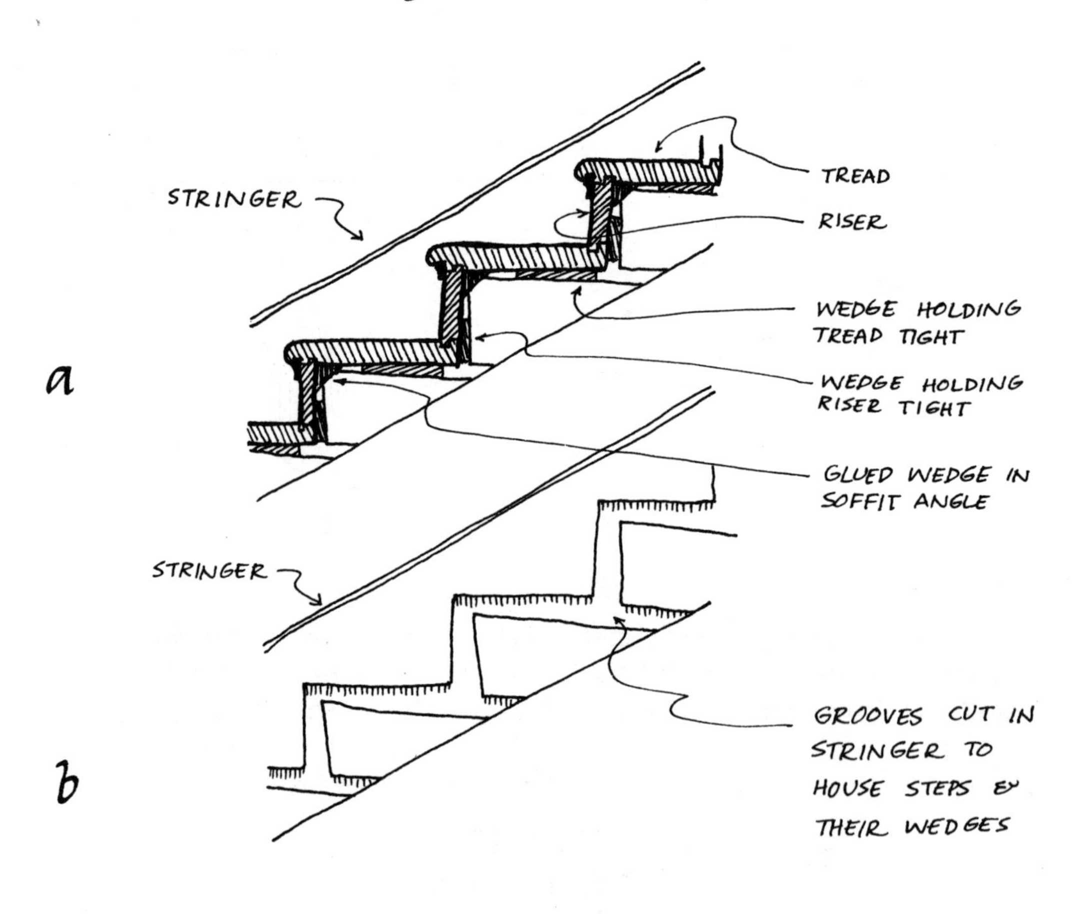

CLOSE STRINGER

A stringer of the type shown opposite is known as a **CLOSE STRINGER**. In a flight of stairs with one string against a wall and the other exposed to view, the former is known as the **WALL STRINGER** and the latter as the **OUTER STRINGER**.

In contradistinction to the close stringer shown opposite, an **OPEN** or **CUT STRINGER** is cut away to receive the treads and risers. The riser and the stringer should be mitered together, the stringer therefore sometimes being known as a **CUT AND MITERED STRINGER**.

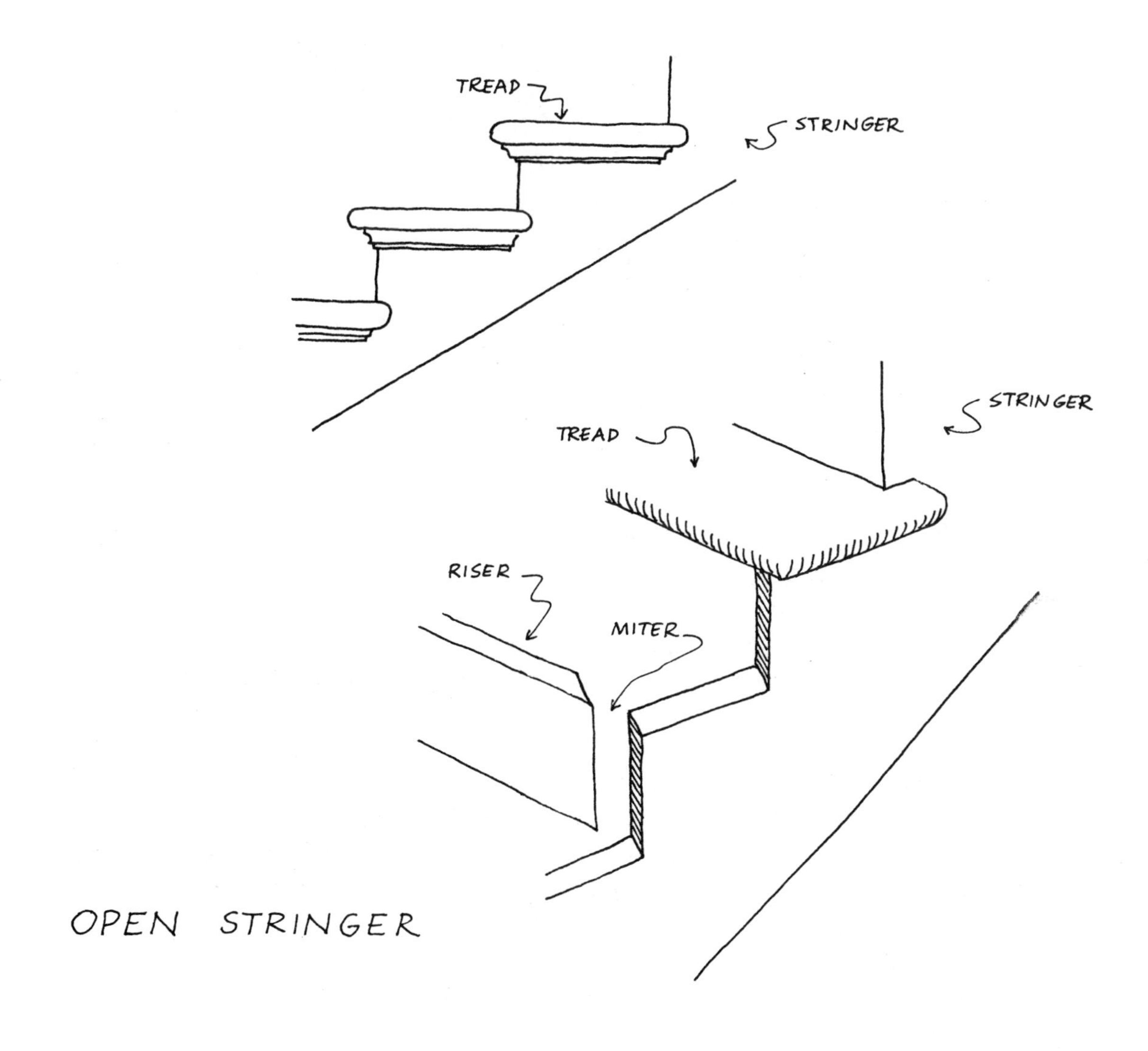

OPEN STRINGER

The nosing of the tread of an open stringer is "returned," that is, carried on around the end, as shown on the previous page. Older types of open stringers often had ornamental brackets fixed under the end nosings of the treads; the stringer was then said to be **BRACKETED**.

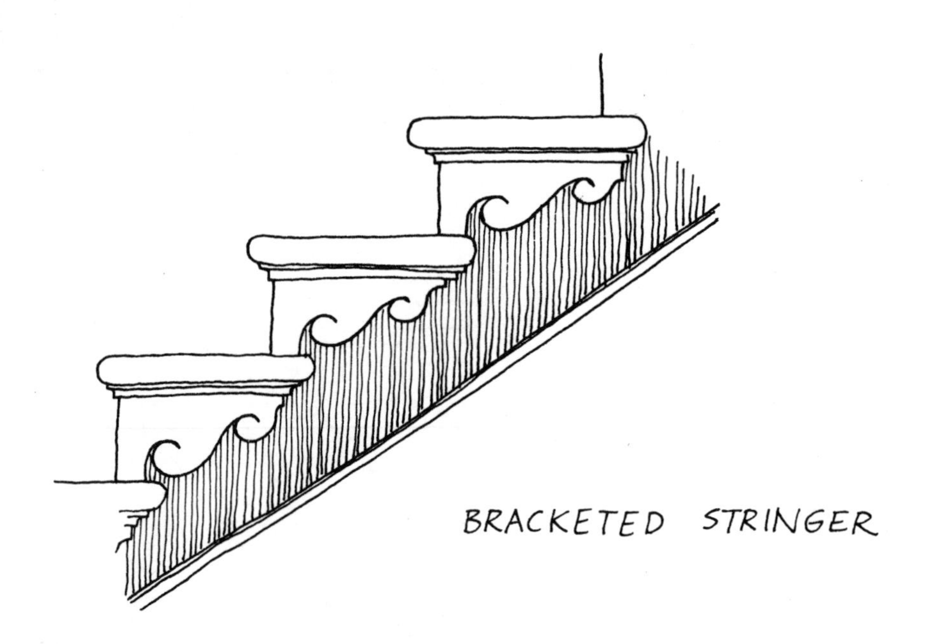

BRACKETED STRINGER

Stringers may be straight or curved; curved stringers are known as **WREATHED STRINGERS**.

A **LANDING** is the floor space at the head of a series of stairs. Such a series is known as a **FLIGHT** or a **FLIGHT OF STAIRS**. (A very descriptive and rather poetic term.)

When the stairs from one floor to another are divided into two flights, the stairs are said to be a **PAIR OF STAIRS**. If the upper flight is parallel to the lower flight, but rises in the opposite direction, a level floor the width of the two flights must be formed; this is known as a **HALF-SPACE LANDING**.

If the upper flight is at right angles to the lower,

or if there are winders at the head of the lower flight or the foot of the upper flight, a landing half the size is sufficient, and this is called a **QUARTER-SPACE LANDING**.

The **RUN** or **GOING** of a flight of stairs is the horizontal distance from the **FOOT** or lowest riser to a point exactly below the **HEAD** or highest riser.

The **RISE** or **HEIGHT** is the vertical distance from floor to landing.

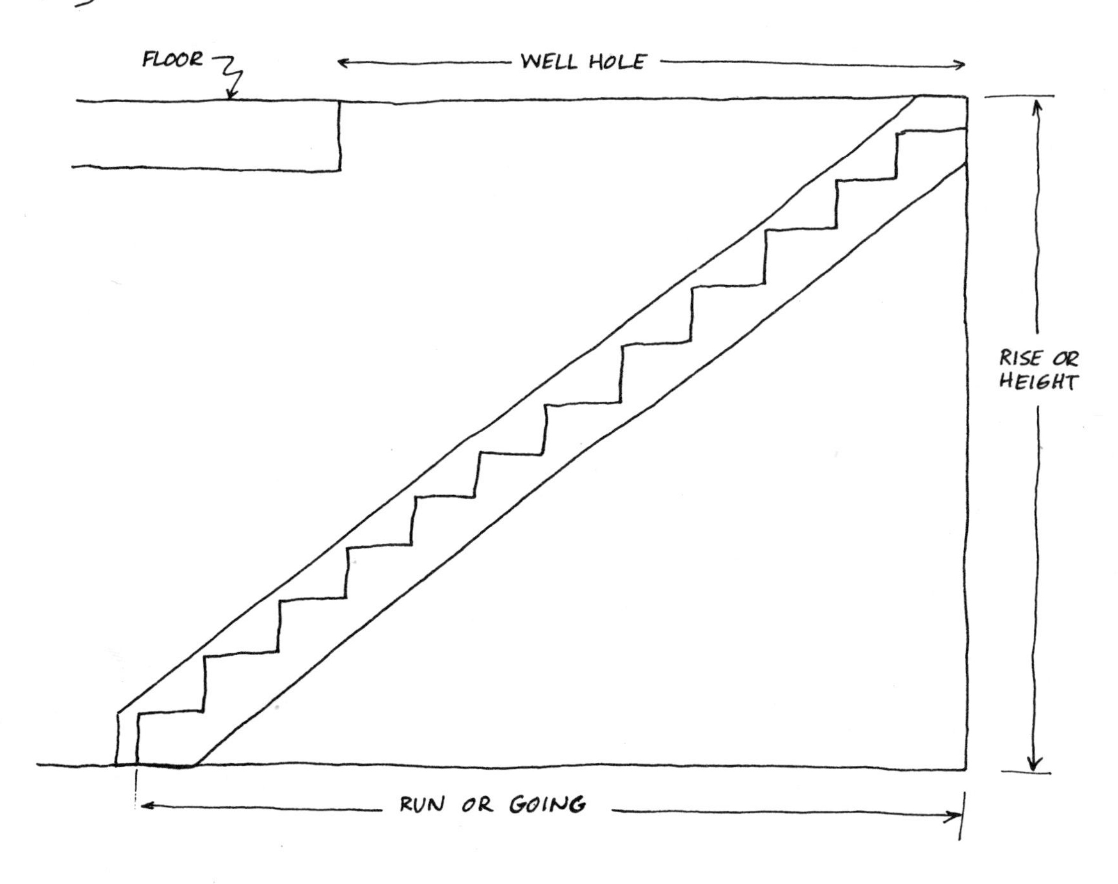

The **WELL** is the place occupied by a flight of stairs, and the **WELL HOLE** is the opening in the floor at the top of a flight of stairs.

NEWELS are posts receiving the ends of stringers and handrails, and are also pillars receiving the treads in a circular staircase. In more elaborate staircases, newel posts are the occasion for much turning and other decorative work, but they function as well if they are simple unadorned posts.

A **DROP** is the lower end of a newel that hangs down at the top of a flight of stairs.

BANISTER is the now more common form and corruption of **BALUSTER**, a word which was originally Italian, meaning "blossom of the wild pomegranate," the shape of which resembled the short pillar, circular in section, slender above and bulging below, used as one of a series to support the **HANDRAIL**, banisters and handrail together forming the **BALUSTRADE**.

HANDRAILS are plain or moulded rails fixed to the walls or newels for convenience and protection.

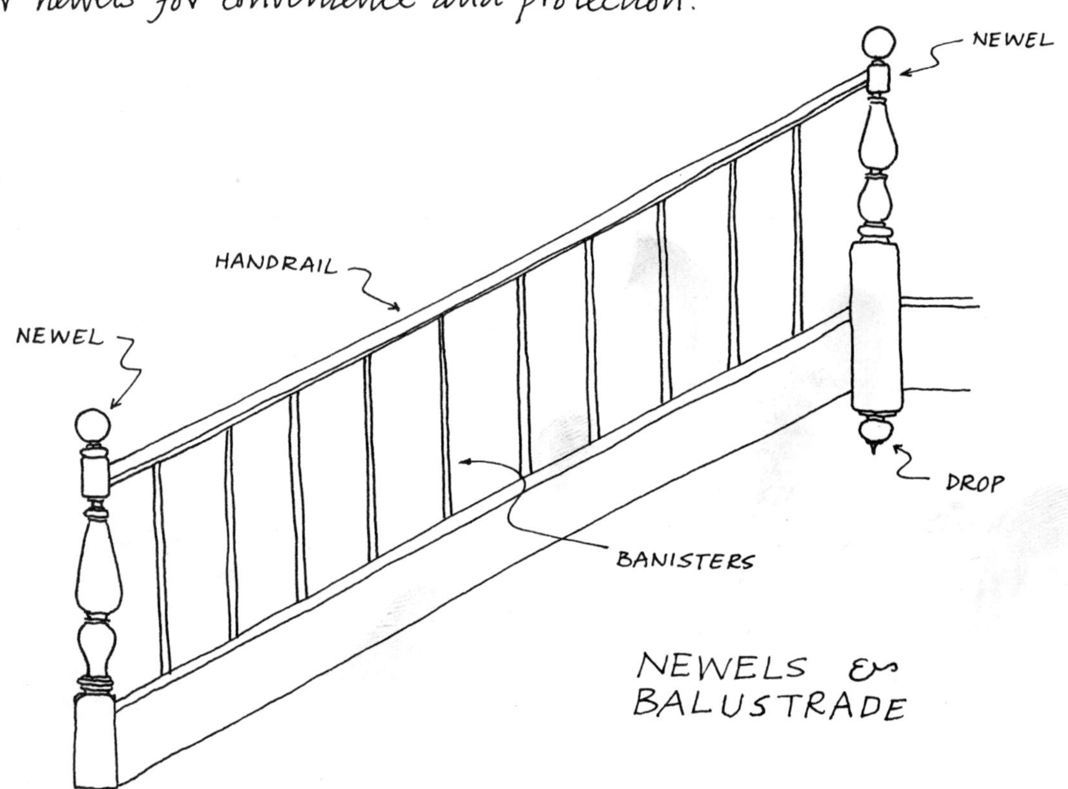

NEWELS & BALUSTRADE

DOG-LEGGED STAIRS are those pairs of stairs where the outer stringer of the upper flight is directly over the outer stringer of the lower flight; both stringers are fixed into the same newel.

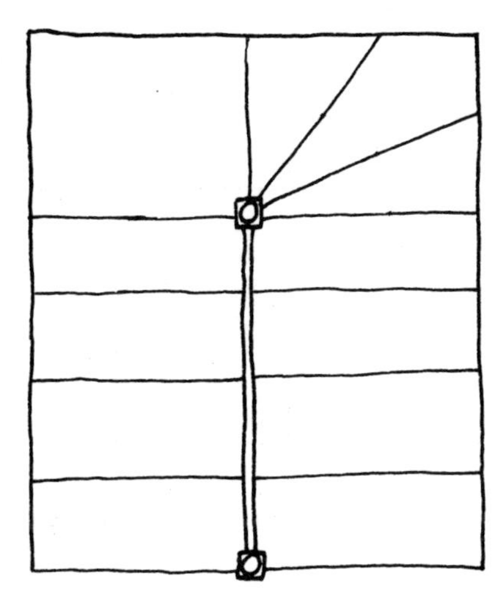

PLAN OF DOG-LEGGED STAIRS

GEOMETRICAL STAIRS have a space or well between the flights and no newel, the stringer being continued in a curve. In this case the stringer and the handrail are said to be **WREATHED**.

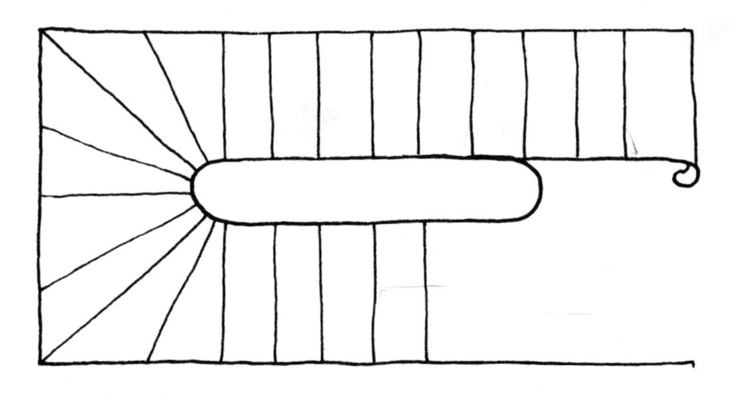

PLAN OF GEOMETRICAL STAIR

Strictly speaking, geometrical stairs not only wind around a central well, but have no half-space or quarter-space landings; stairs that do are properly called **WELL STAIRS**.

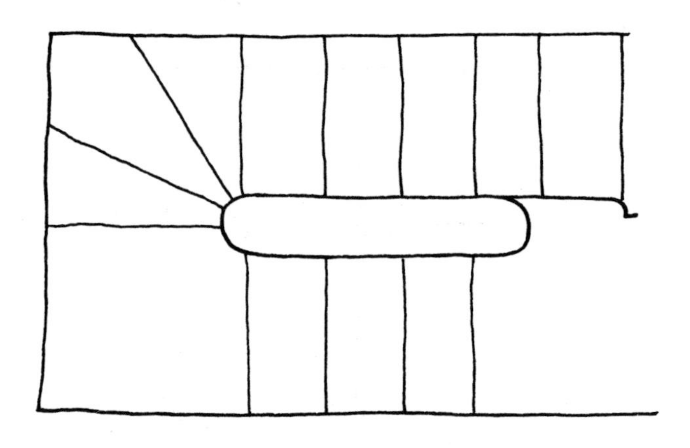

PLAN OF WELL STAIRS

CIRCULAR STAIRS have the treads and risers framed into a central post or newel around which the steps wind.

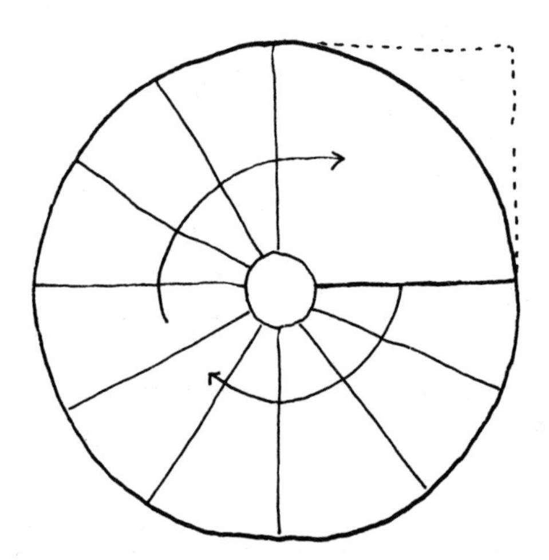

PLAN OF CIRCULAR STAIRS

Other more complicated forms of stairs are **COCKLE STAIRS**, which is a winding staircase, a **SPIRAL STAIRCASE** in which all the steps are winders but revolve around a well, and **ELLIPTIC STAIRS** in which each tread converges so that in plan an elliptic ring is formed.

The **CARRIAGE** is the framework, including the stringers, which supports the steps.

BEARERS are supports for winders wedged into the walls and secured by the stringers.

TRIMMERS are the beams to which stringers are fixed at the top of one stair and the start of the next.

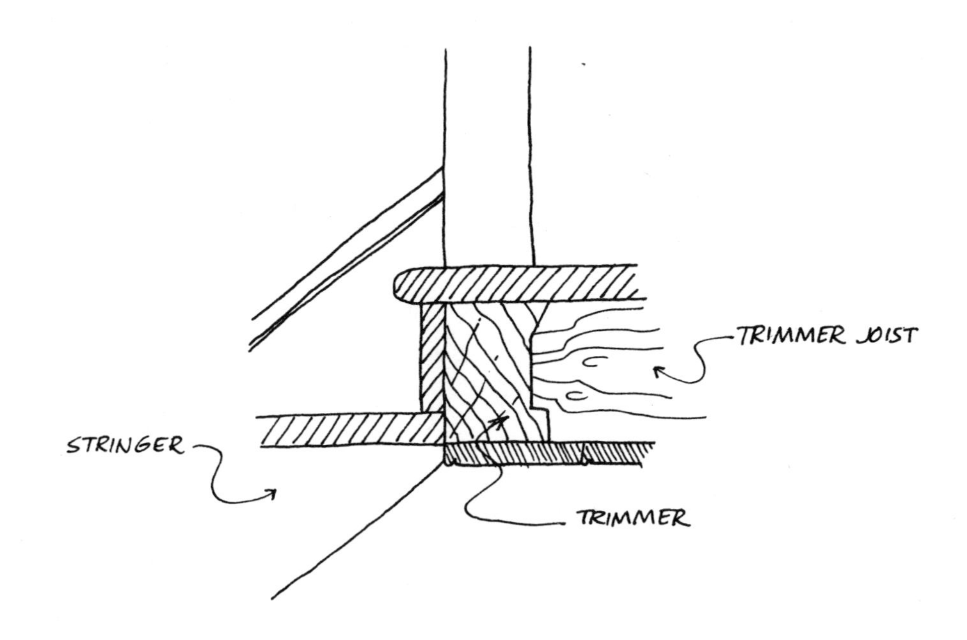

Where possible, stairs are considered best made with a right-hand wheel rather than left-hand, but it should be remembered that the form of stair can be worked out better in new buildings than in old buildings being remodeled.

SIMPLE STAIRS

Before proceeding to the complexities of orthodox stairbuilding, some simpler methods are here described which may be found useful for porch or attic steps, or for temporary purposes.

Next to the common ladder, the simplest kind of stair is a thick plank placed at a convenient slope to form the ascent, with pieces of wood nailed to it to give a firm footing.

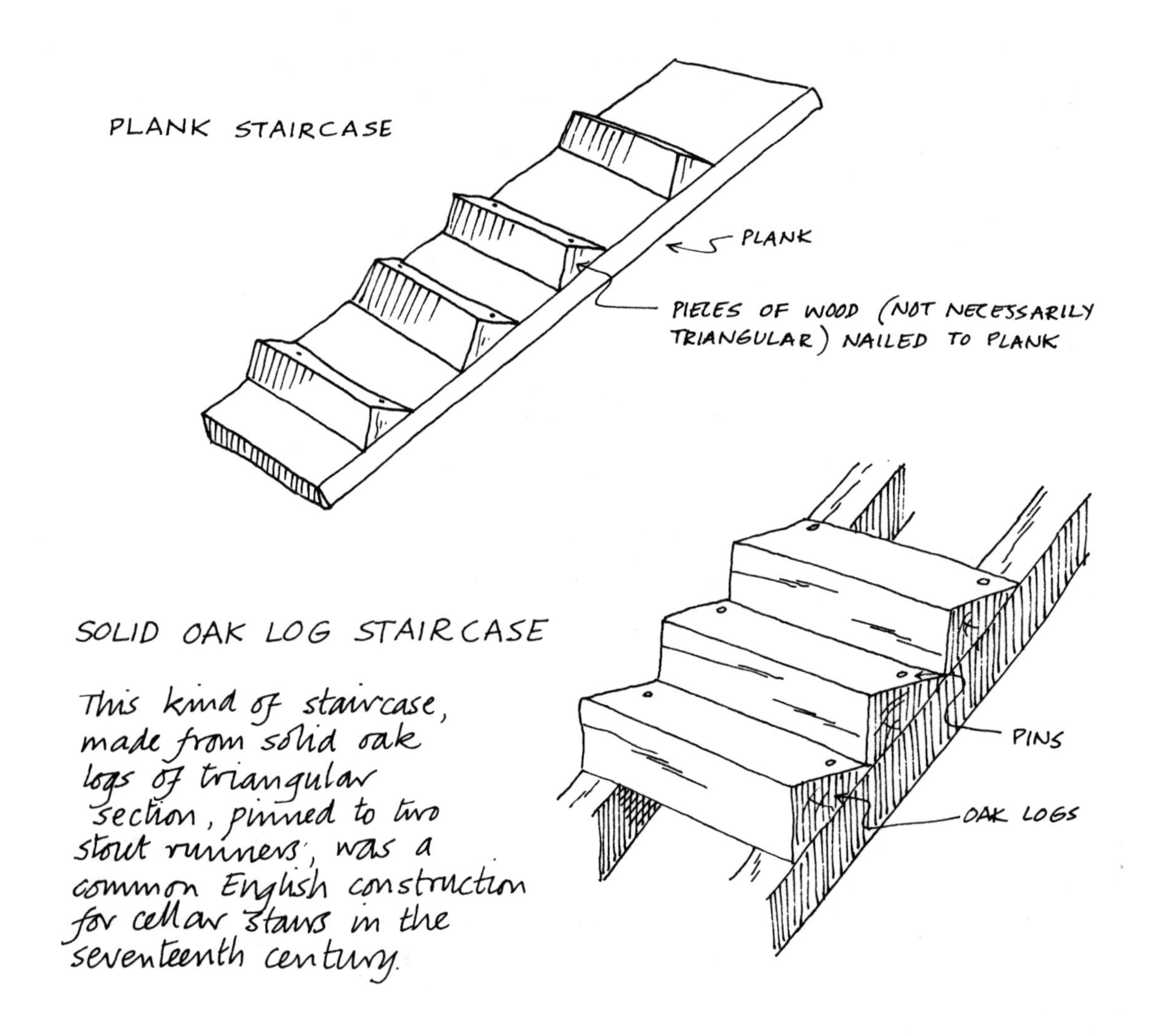

PLANK STAIRCASE

SOLID OAK LOG STAIRCASE

This kind of staircase, made from solid oak logs of triangular section, pinned to two stout runners, was a common English construction for cellar stairs in the seventeenth century.

Another form of simple stair which is of interest because of its antiquity is shown below. While it doesn't require much accuracy or skill, there is, however, a considerable amount of labor involved.

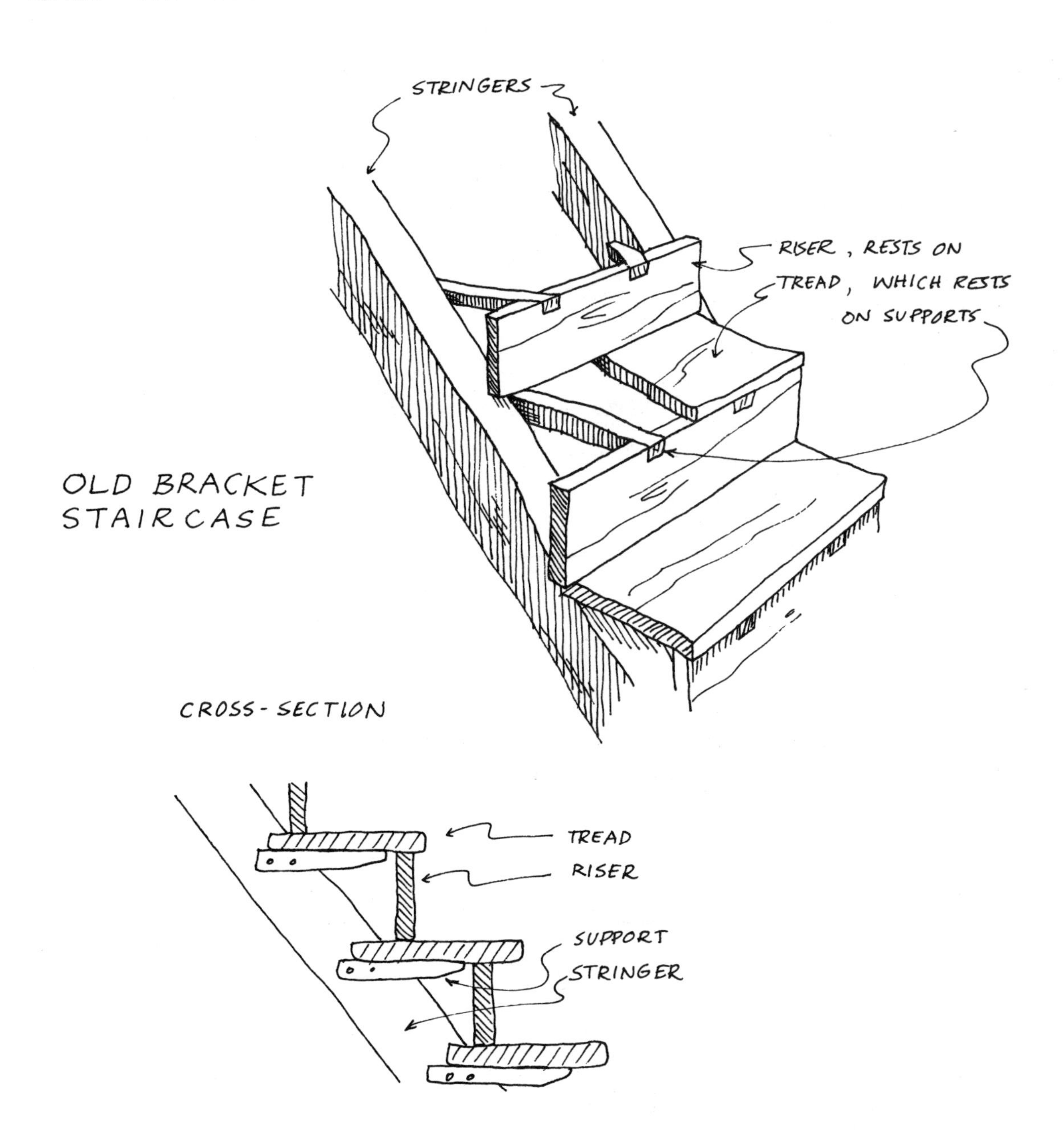

The stair next in degree of complexity to the plank stair is one composed of boards forming treads wide enough to provide a secure footing, fixed to two raking boards forming the stringers.

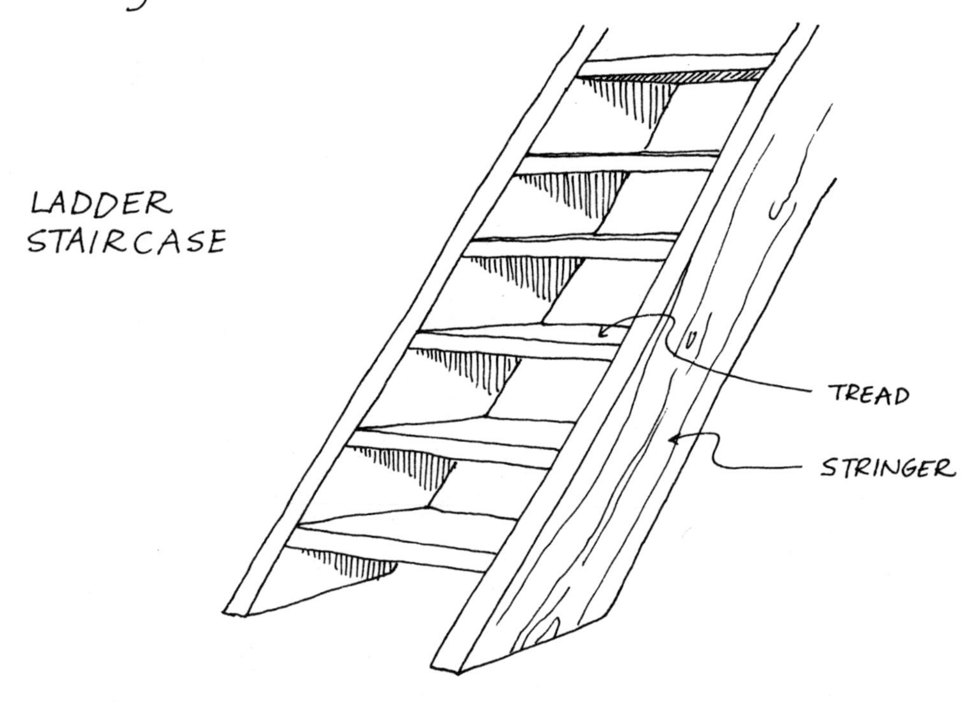

The treads may be nailed directly to the stringers; this is the quickest and poorest solution.

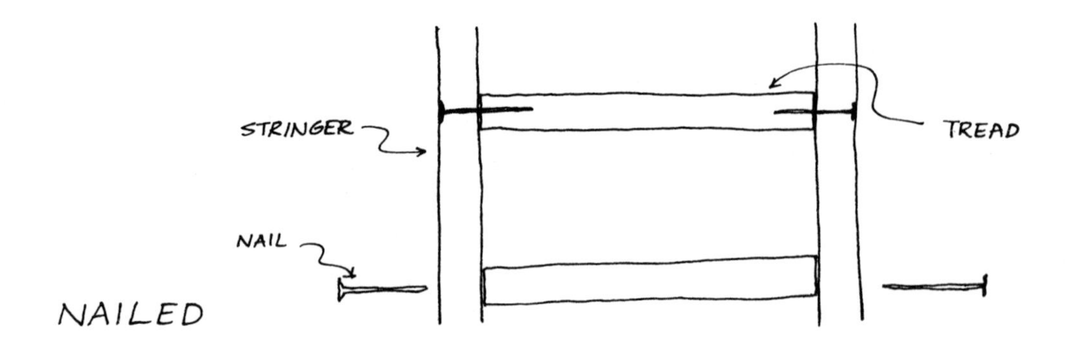

A better method than simply nailing through the stringers into the ends of the treads is to house the treads into the stringers, and then nail or screw.

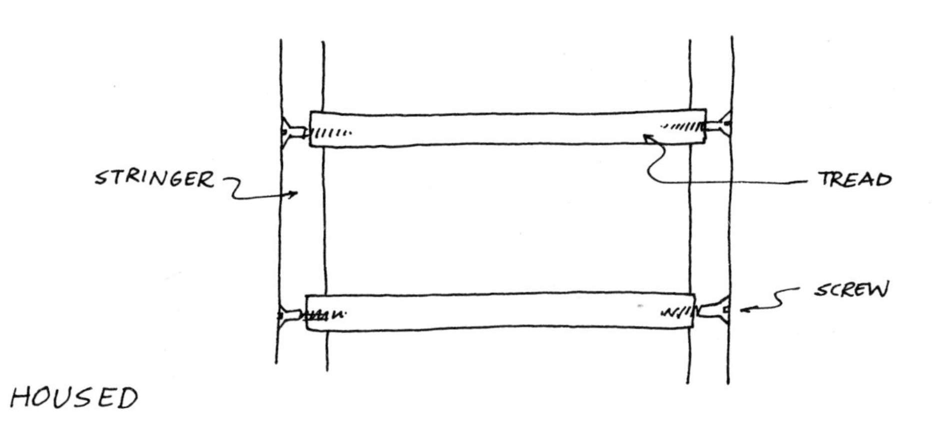

Even better in terms of secureness, though perhaps not so neat in appearance, is the use of cleats. These are small strips fixed to the inside of the stringers; they may be nailed or screwed, though screws are stronger. To these are similarly nailed or screwed the treads. This system is perhaps easier than rabbeting out stringers for housing joints and also affords greater lateral strength.

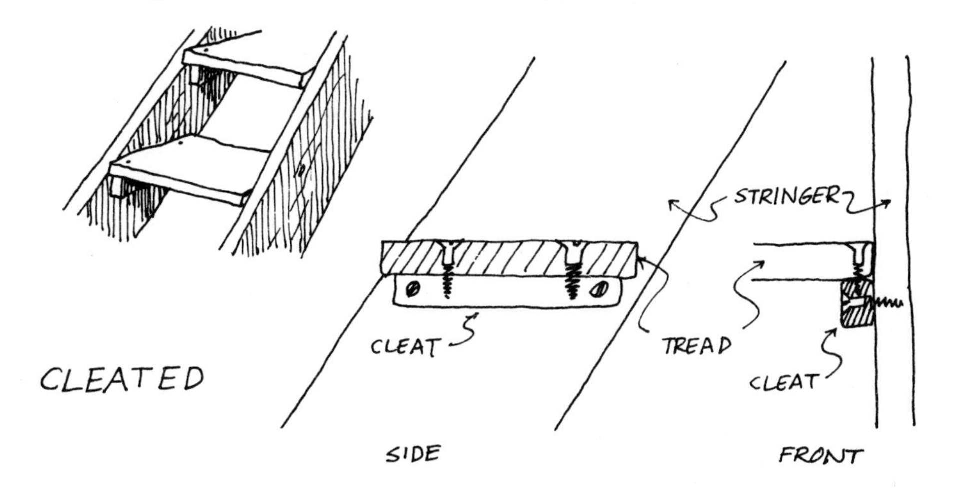

Yet another method of constructing ladder stairs is to mortise the treads into the stringers. The mortises are sometimes rectangular, as at "a," and sometimes for a nicer effect the mortises may follow the slope of the stringers, as at "b." For "Mission Style" staircases the tenons are carried right through the stringers and secured by keys, as at "c."

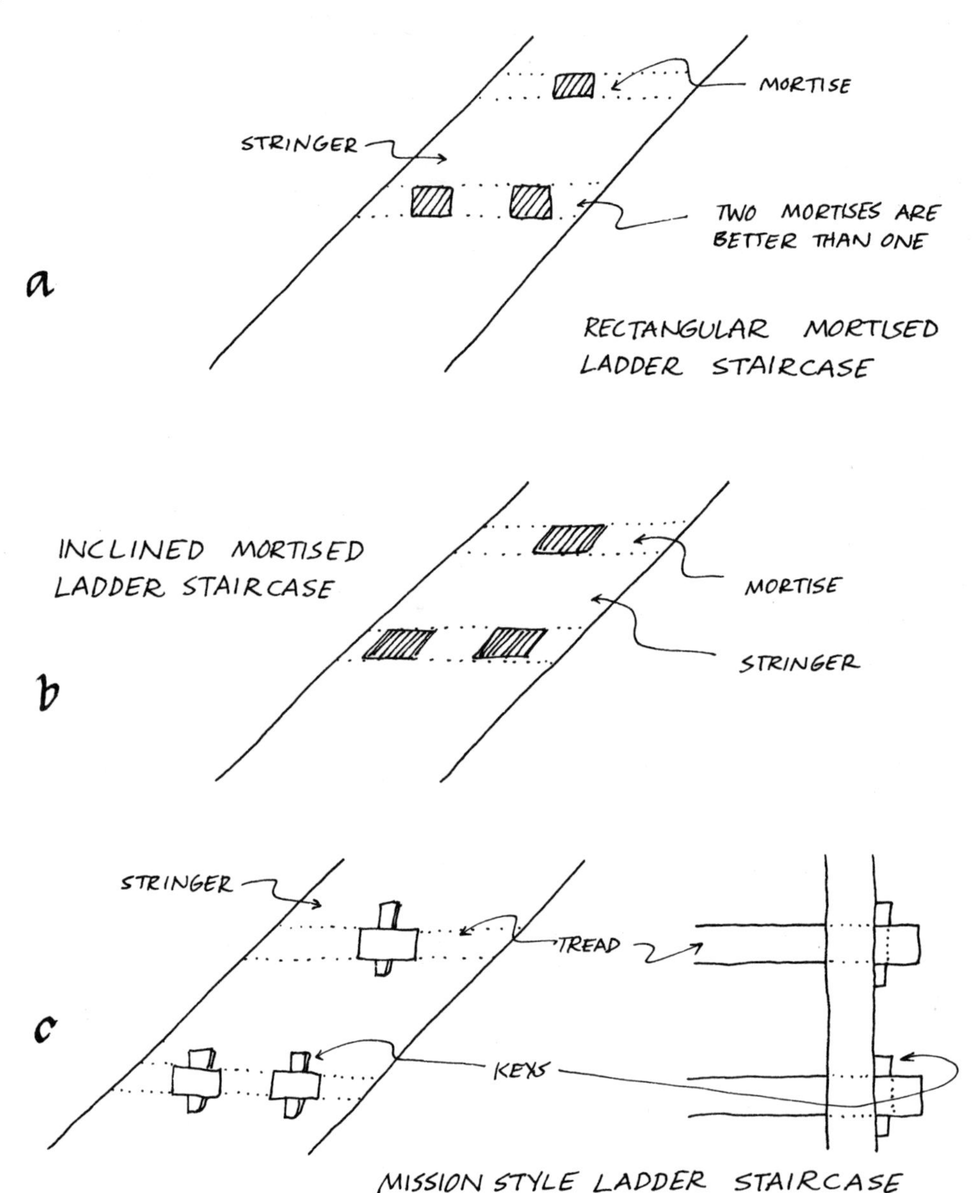

The stringers are sometimes bound together by rods which are secured on the outside faces with washers and nuts. Such rods should be placed near the middle of a step, and close to its under side.

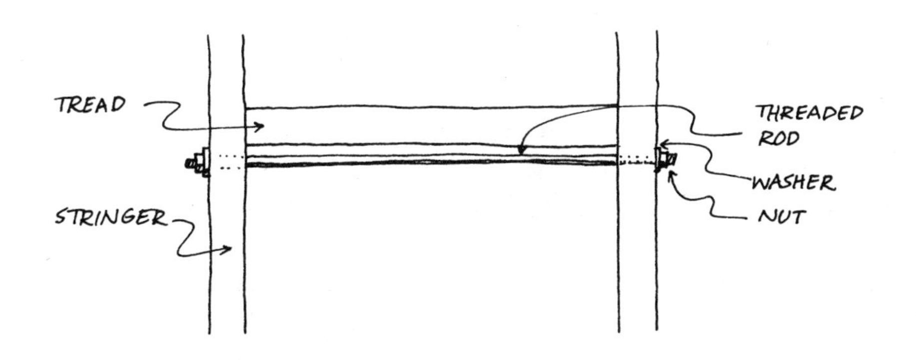

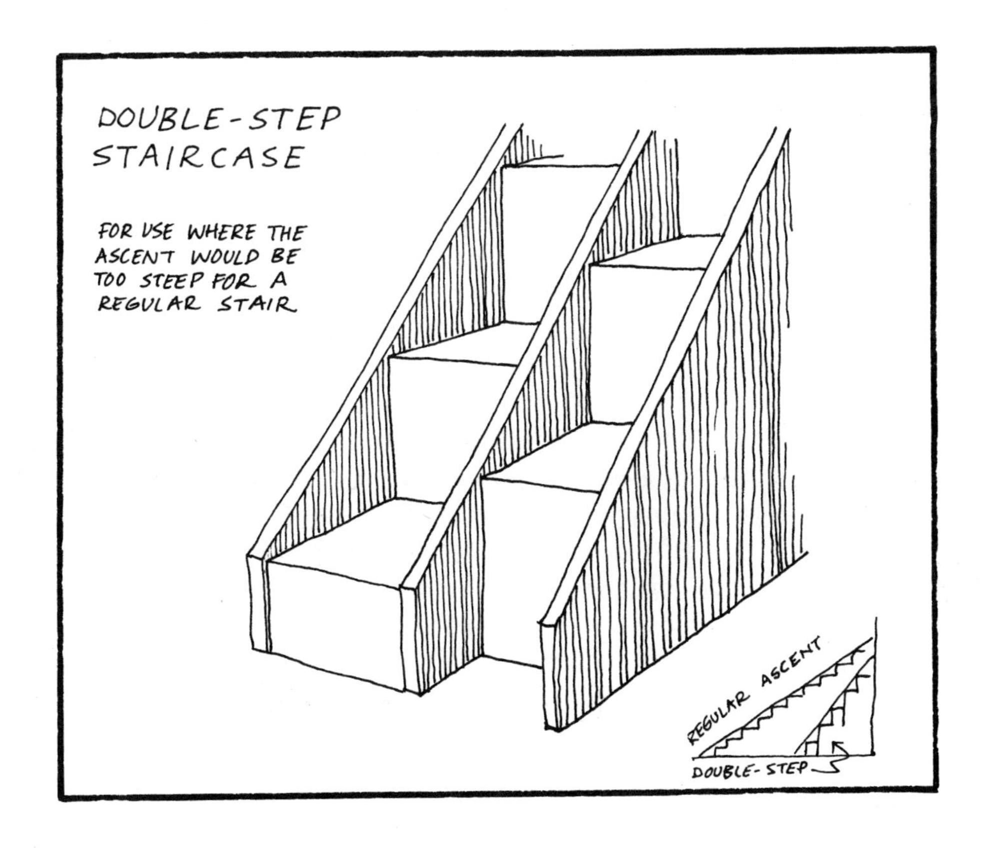

DOUBLE-STEP STAIRCASE
FOR USE WHERE THE ASCENT WOULD BE TOO STEEP FOR A REGULAR STAIR
REGULAR ASCENT
DOUBLE-STEP

PROPORTIONS

The two most important aspects of good stair design are that there be sufficient headroom and that the stairs be comfortable to ascend and descend. This last is governed by the proportions of the treads to the risers.

It is a general maxim that the greater the breadth of a step the less should be the height of the riser; and conversely, the less the breadth, the greater should be the height.

Experience shows that a step with a tread of 12" and a riser of $5\frac{1}{2}$" may be used as a comfortable standard. Therefore, by using the following formula, we may determine the breadth of the tread if we know the riser, or the height of the riser if we know the tread:

Let **T** be the tread and **R** the riser of any step which has the proper proportions; then to find the proportion of any other tread **t** and riser **r**:

$$\frac{R \times T}{r} = t, \text{ or}$$

$$\frac{T \times R}{t} = r$$

For example, taking a tread of 12" and a riser of $5\frac{1}{2}$" as the standard, then, to find the breadth of the tread when the given riser is 8", and substituting these values for **t** and **r** in the formula, we have:

$$\frac{12 \times 5\frac{1}{2}}{8} = 8\frac{1}{4} \text{ inches as the breadth of the tread.}$$

Another example, to find the height of the riser when the breadth of the tread is known to be 13":

$$\frac{T \times R}{t} = r$$

$$\frac{12 \times 5\frac{1}{2}}{13} = 5\frac{1}{13}"$$ as the height of the riser.

This proportional variation between tread and riser may also be illustrated graphically, which can be useful if you want to make any drawings before you start to build (which often helps).

Firstly, substituting, for the sake of working with whole numbers, the dimensions in so many half inches, and accepting again the comfortable proportion of 12 to $5\frac{1}{2}$, draw a right-angled triangle with a base of **24** and a perpendicular of **11**.

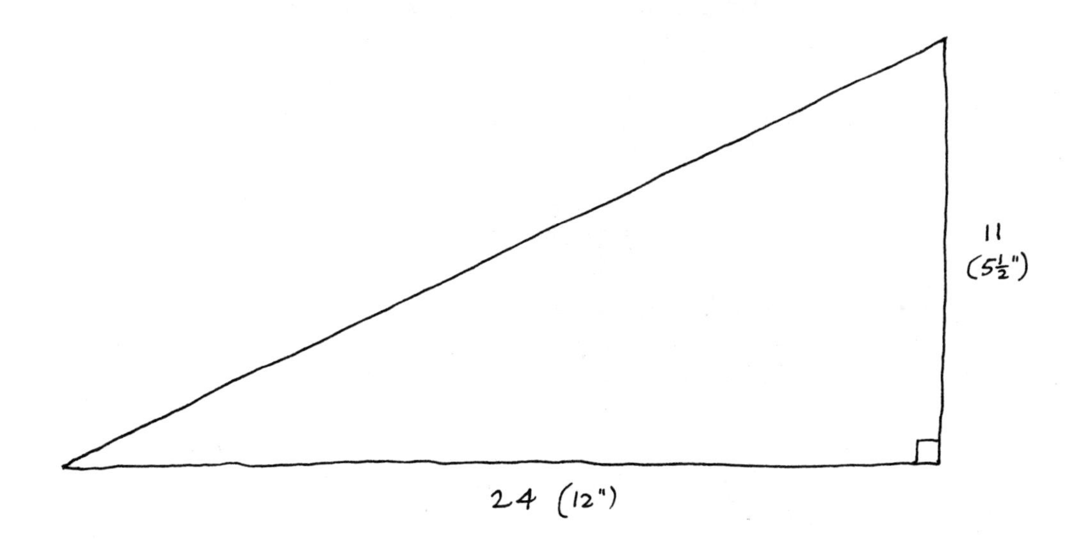

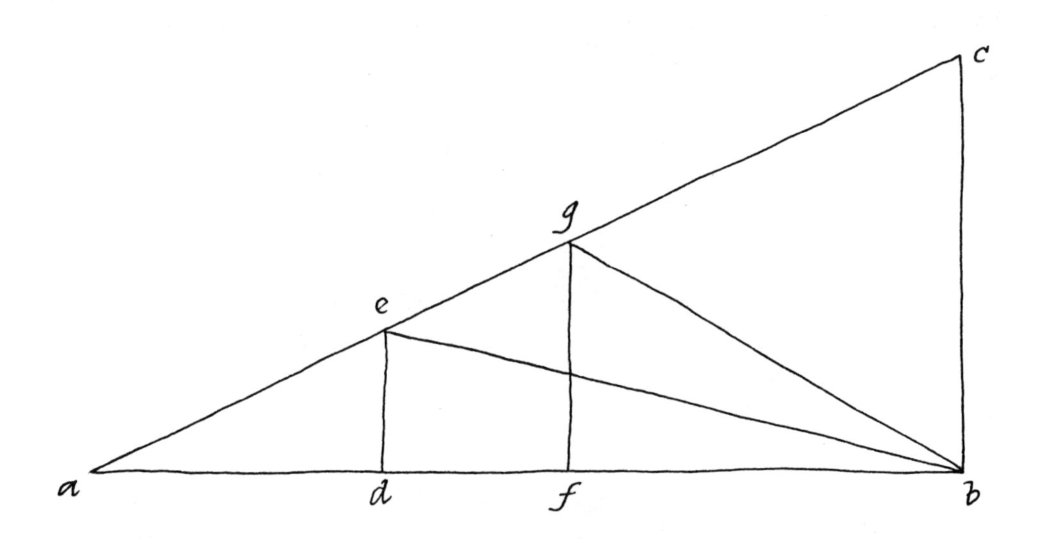

In this right-angled triangle, which we shall call a b c, ab equals **24** and bc equals **11**.

To find the riser corresponding to a given tread, from b set off on ab the length of the tread, which we shall call bd.

Through d draw the perpendicular de, meeting the hypotenuse in e; then de is the height of the riser, and if we join be, we get the slope of the ascent.

Similarly, where bf is the width of the tread, fg is the riser, and bg the slope of the stair.

Another method is by use of the following table, which, once again, accepts as the standard the proportions of 12" to $5\frac{1}{2}$" and extends the treads in inches and the risers in half inches.

It can be easily seen that these treads and risers will suitably pair together.

TREADS IN INCHES	RISERS IN INCHES
6	$8\frac{1}{2}$
7	8
8	$7\frac{1}{2}$
9	7
10	$6\frac{1}{2}$
11	6
12	$5\frac{1}{2}$
13	5
14	$4\frac{1}{2}$
15	4
16	$3\frac{1}{2}$
17	3
18	$2\frac{1}{2}$
19	2

Another very useful rule may be expressed as follows:

WIDTH OF TREAD + TWICE THE HEIGHT OF RISER = 24, or:

$$T + 2R = 24$$

Transposing this we get:

$$T = 24 - 2R \text{ , and}$$

$$R = 12 - \frac{T}{2}$$

Here are two examples showing the application of this rule:

First, let the riser be 7", then $T = 24 - (2 \times 7) = 24 - 14 = 10"$.

Second, let the tread be 12", then $R = 12 - \frac{T}{2} = 12 - 6 = 6"$.

To avoid confusion the above examples have been expressed entirely in inches, but they work equally well if transposed into the metric system. To arrive at millimeters from inches, multiply the inches by **25.4000508**.

Since such multiplication can prove tiresome (if not inaccurate and productive of unhandy results), the following metric methods may be used instead.

Based on the assumption that the average length of a step is **600 mm** and that it is twice as tiring to climb upward as it is to walk forward, use the following formulae to obtain a comfortable riser and tread:

1. ONE TREAD + TWO RISERS = **600 to 620 mm**.

Example: tread **250 mm** + twice riser = **610 mm**
twice riser = **610 − 250 = 360 mm**
riser = $\frac{36}{2}$ = **180 mm**.

2. RISER × TREAD = **450**

Example: tread **250 mm** × riser = **450**
riser = $\frac{450}{250}$ = **180 mm**.

Despite all of the above, however, it is seldom that the proportion of the tread and the riser is a matter of choice; the room allotted to the stairs usually determines this proportion; but the above rules will be found useful as guidelines, which you should try to approximate.

LAYING OUT STAIRS

Although the design of stairs is properly the work of an architect, the carpenter is often the one who has to do the final figuring out. This is actually the most difficult part and must be done most carefully or disaster will ensue. Once all the measurements have been correctly ascertained, the actual woodwork is not very difficult.

The most common situation the carpenter faces is having to build stairs in an already given space. If the stairs could be made to perfect proportions and then fixed in place with no obstructions, there would be few problems. However, the run or going is often predetermined and the problem is to figure the slope and size and number of the steps.

The proportions of tread to riser have been discussed in the previous section. The other important consideration when figuring stair layout is headroom.

Since the given dimensions are variable, i.e., the opening in the second floor may be fixed, the available run may be fixed, the problem may be solved from different directions; but an example from one direction should suffice to explain the principles, and the variables may then be substituted as the situation dictates.

Let us therefore assume the distance between floors, i.e., the rise for a straight flight of stairs to be "x," and the opening in the second floor to be "y."

The first step is to draw two lines representing the two floors. Call these two lines ab and cd, as shown over.

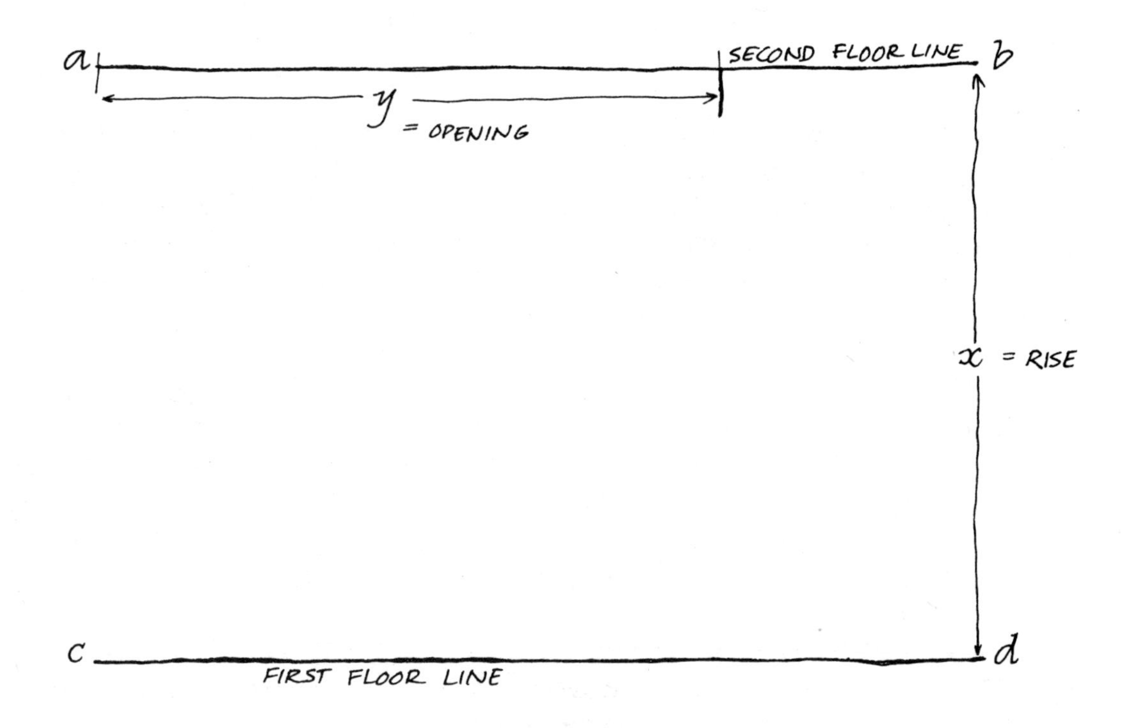

Now measure down from the second floor line to the ceiling underneath it – this will include the thickness of the finish floor, the subfloor if any, the joists, and the finished ceiling.

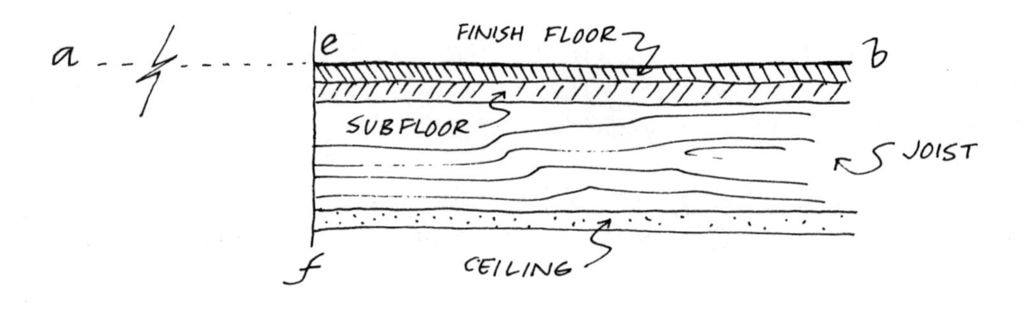

Call this measurement ef and mark it on your plan.

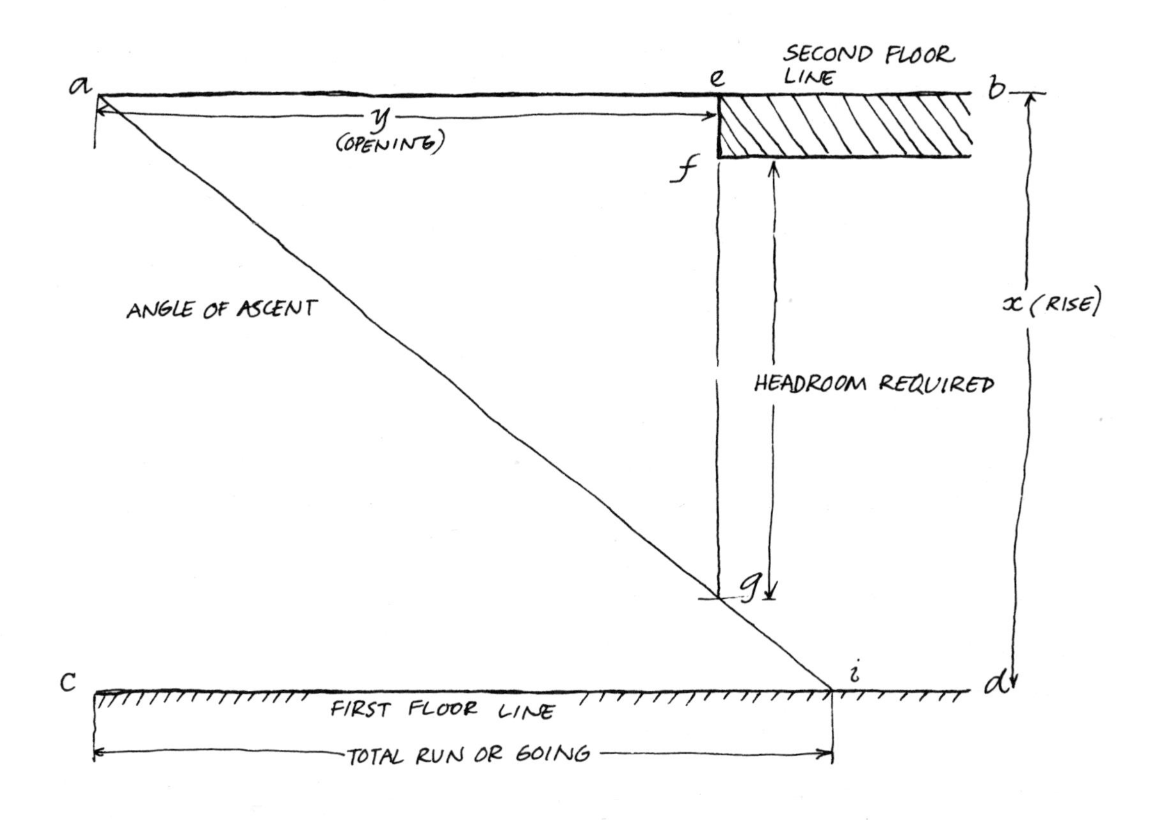

Now draw a vertical line down from f to represent the amount of headroom required to g.

Next draw a line from a (the forward edge of the second floor opening) through g to meet the first floor line ab. Call the point where this line meets ab "i." The line ai is the angle of ascent and ci is the run or going.

The amount of headroom depends on what is convenient. If you make it about **7' (2.13 m)** it will sufficiently accommodate most human beings – except for giants, and they are usually accustomed to bending in buildings anyway.

You now know the rise (let us assume 9' (2.74m)) and the run. The next step is to find out how many steps will fit in this staircase.

Take as an average a riser of 7" and a tread of 10". (Remember the formula: 2 × RISER + TREAD = 24; 2 × 7 + 10 = 24.) Simply divide the total rise of 9' by the height of an average riser of 7". The answer is:

(9 × 12) (to convert to inches) ÷ 7 = 15.4.

Since all risers must be the same size in any one staircase or you will trip over the odd one, we cannot have any fractions, so assume the number of risers to be 15.

There are two ways to discover the actual measurement of a riser now. Either divide the total rise by the number of risers:

(9 × 12) ÷ 15 = 7.2", or

take a story pole (a long straight stick) and, setting the dividers at a little over 7", step off the 9' distance representing the total rise. If there is a remainder, adjust the dividers until the 9' is exactly divided into 15 parts. Measure one of these parts and that is the measurement of a riser.

Now measure the total run as found in the diagram on the previous page as ci. Let us assume that this works out to be 11'4½". Bearing in mind that the number of treads equals the number of risers minus one, as shown on the next page, we can work out the measurement of a tread by using the following formula:

TREAD = RUN ÷ NUMBER OF STEPS

TREAD = 11'4½" ÷ (15 − 1)

TREAD = 136.5 ÷ 14 = 9.75"

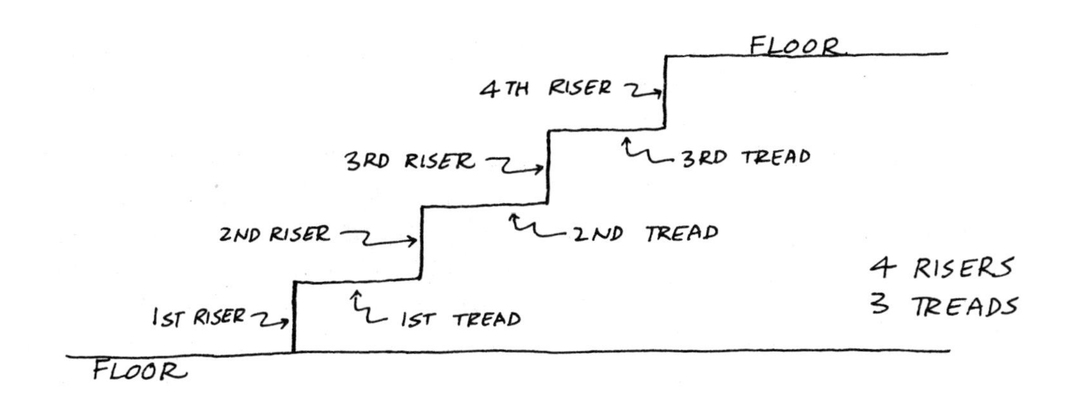

SHOWING NUMBER OF TREADS EQUALS NUMBER OF RISERS MINUS ONE

Now that you know the actual measurements of the treads and risers, complete the drawing as shown below.

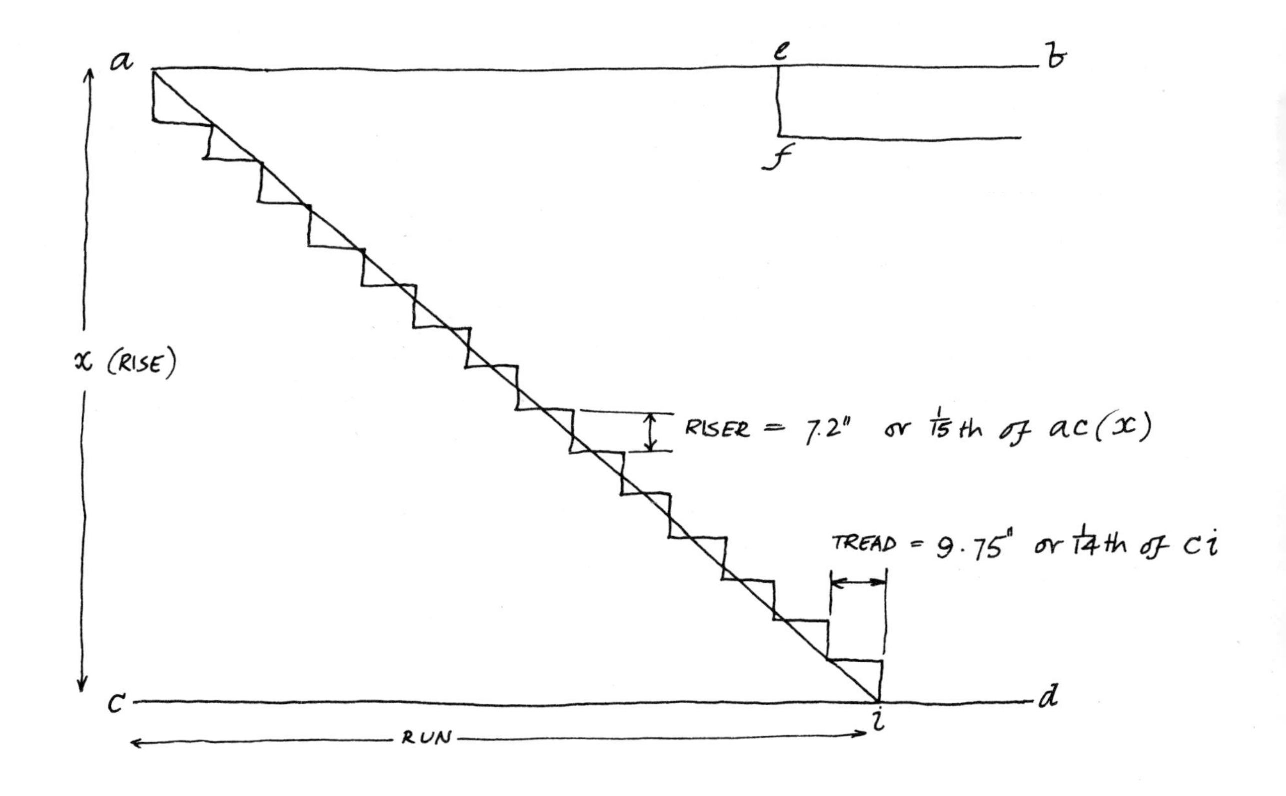

Note that the line of nosings, if you were to draw it in, does not coincide with the previously drawn angle of ascent.

There are two ways to complete the drawing on the previous page: one is to measure up from i **7.2"** vertically, and then measure in horizontally **9.75"** until you reach the top; or, more easily, divide x into **15** parts and draw a series of horizontal lines across the plan, and then divide ci into **14** parts and draw a series of vertical lines to intersect the parallel lines. The intersections thus formed constitute the previously defined line of nosings.

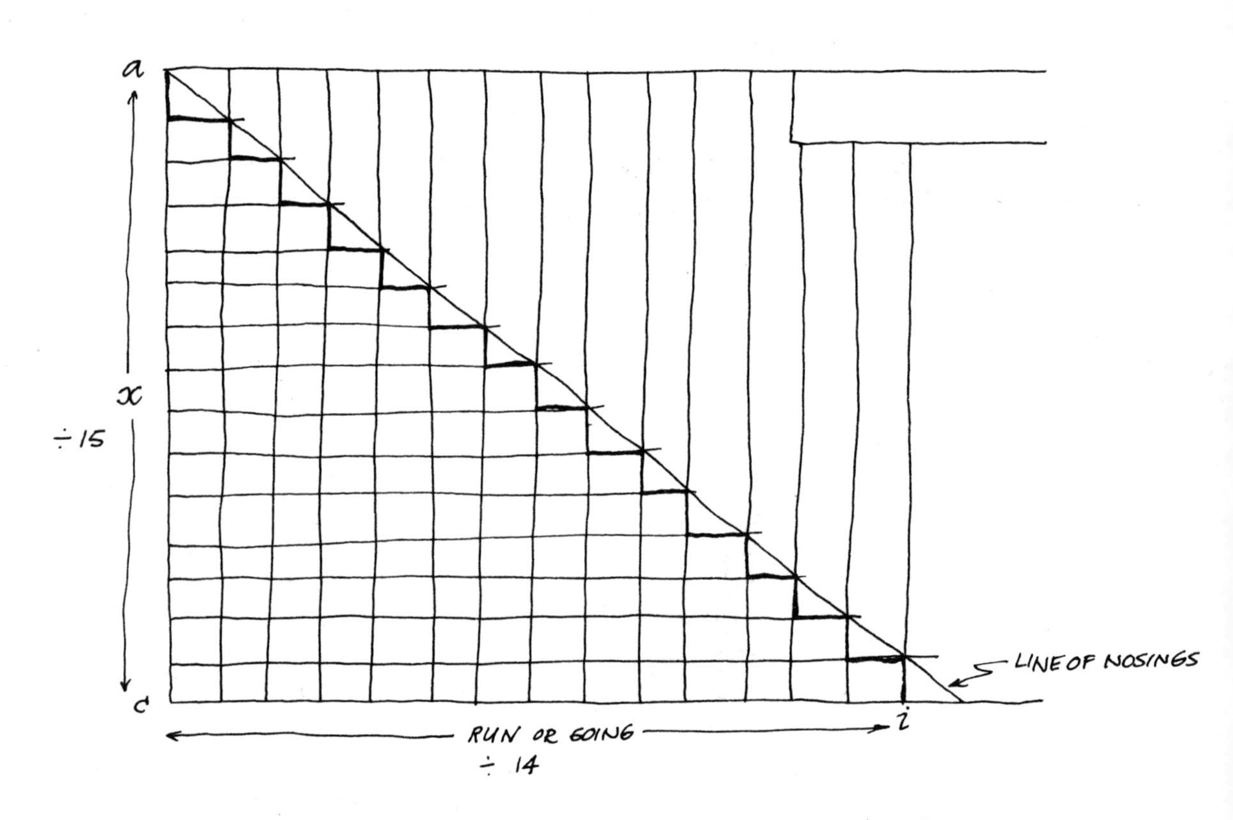

DRAWING IN THE TREADS & RISERS

CUTTING THE STRINGERS

When the drawing has been completed and you have found all the pertinent dimensions, the next step is to make the stringers.

There are, as mentioned briefly before, three main types of stringers: open, close, and bracketed. The close stringer is and has been traditionally superior, since the treads and risers can be wedged in place, and when shrunk and beginning to creak, can be made tight again by simply driving the wedges further in. To prevent open stringers from creaking is a little more difficult, but the construction is much simpler. However, the laying out of all three types is the same in principle, so we shall proceed with the easiest type as our example.

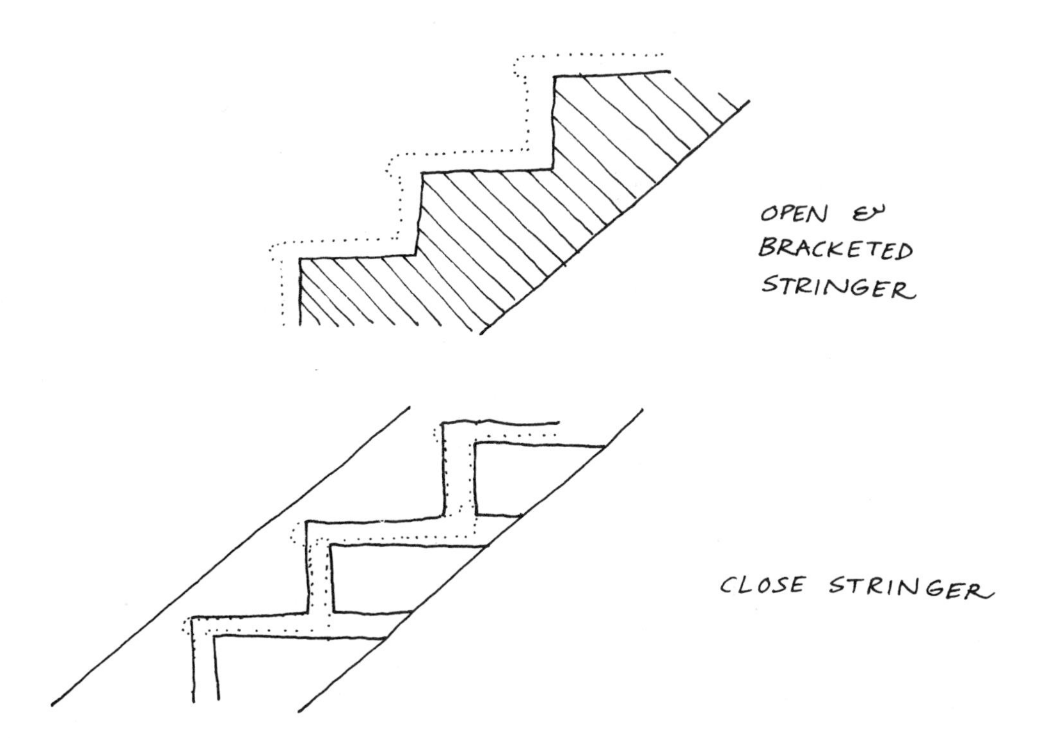

The open stringer may be marked for cutting out by the use of a carpenter's steel square (often made from aluminium to make it lighter), of a steel square with a fence attached, or by the use of a pitch board – this last being the most accurate of all three methods.

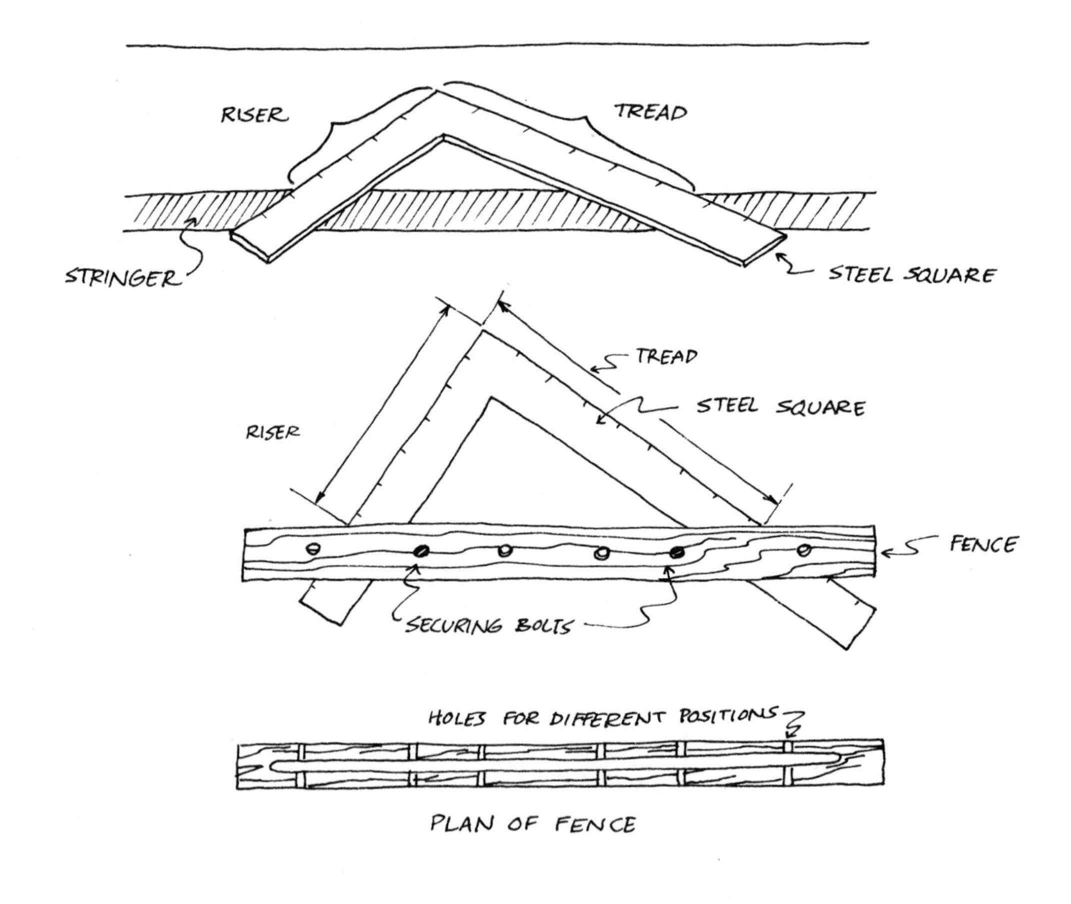

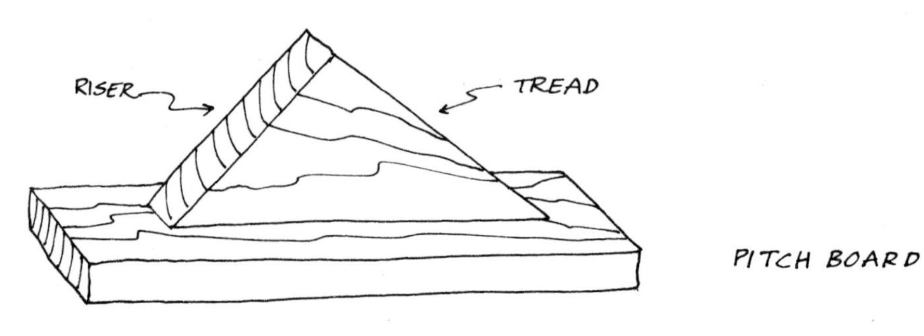

PITCH BOARD

Using the steel square, as shown opposite is all right for common porch, attic, or basement stairs. For a little greater accuracy, make a fence out of a piece of straight hardwood, such as a piece of oak. Simply saw a slot down the middle and bore a few holes for nuts and bolts to clamp the fence tightly on the square. Wing nuts make any adjustment that may be necessary easy to make.

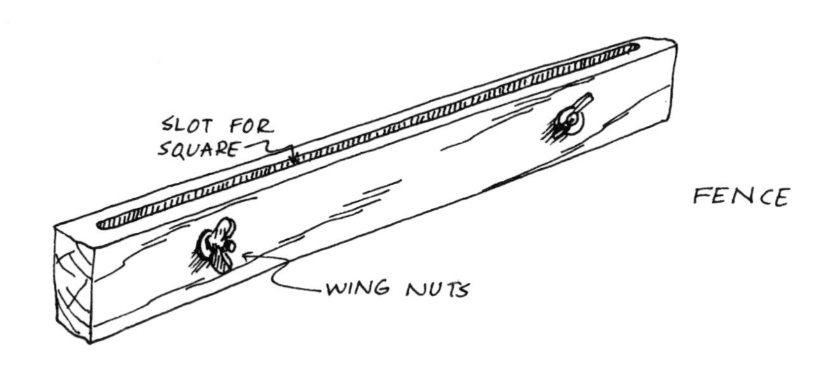

FENCE

The pitch board, which is cut to the exact measurements of the tread and riser, ensures even greater accuracy and only takes a few moments to make.

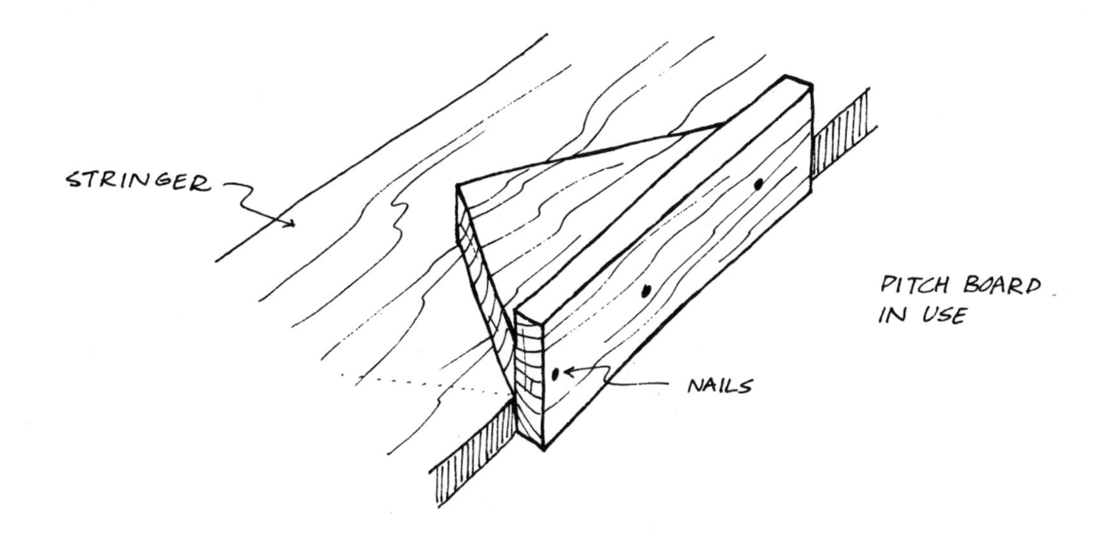

PITCH BOARD IN USE

The most exact method, however, is to use a pair of dividers set to the step length or distance between nosings. This is found by calculating the distance along the face of the stringer and dividing it by as many steps as there will be.

This is a little difficult, since we have as yet no measurement for the finished overall length of the stringer. This can be seen by looking at the drawing below.

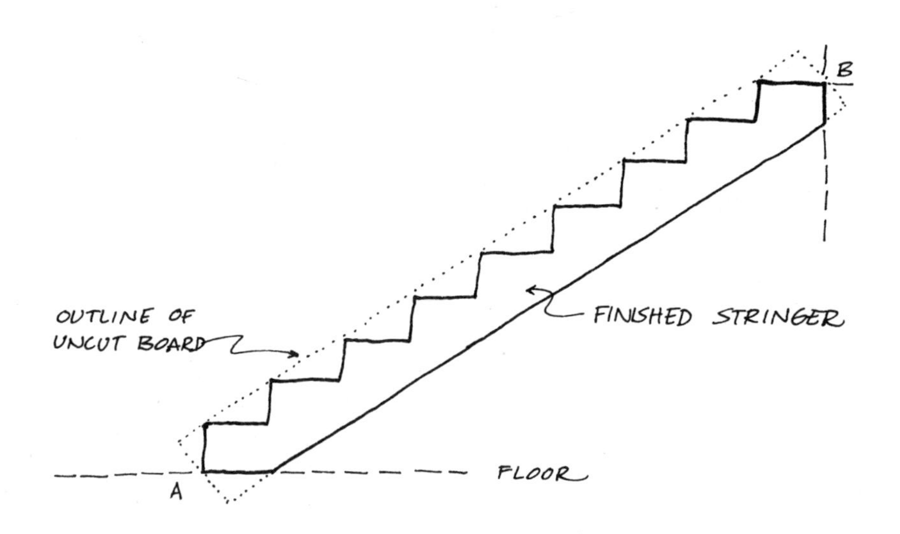

We have not calculated AB, nor do we have to. Since the riser height is known, simply lay off the second step a sufficient way along the board to leave room for the first step, and then, having thereby established the vertical angle of the riser, mark the first riser and draw a line at right angles back to the underside of the stringer.

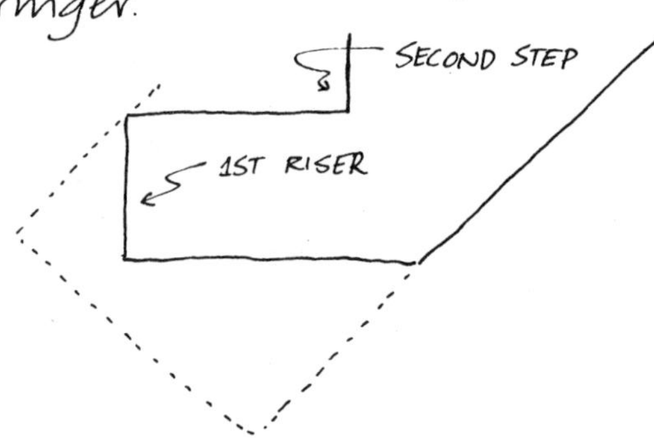

By using the pitch board as many times as there are steps, the point B on the illustration opposite will ultimately be reached. To check this, use Pythagoras' theorem which states that in any right-angled triangle the square of the hypotenuse equals the sum of the squares of the base and the perpendicular.

The run or going equals the base in the above theorem, and the total rise equals the perpendicular, both of which we know.

We need make only one adjustment to calculate the step length, and this is to subtract from the total rise the height of one riser, as shown below.

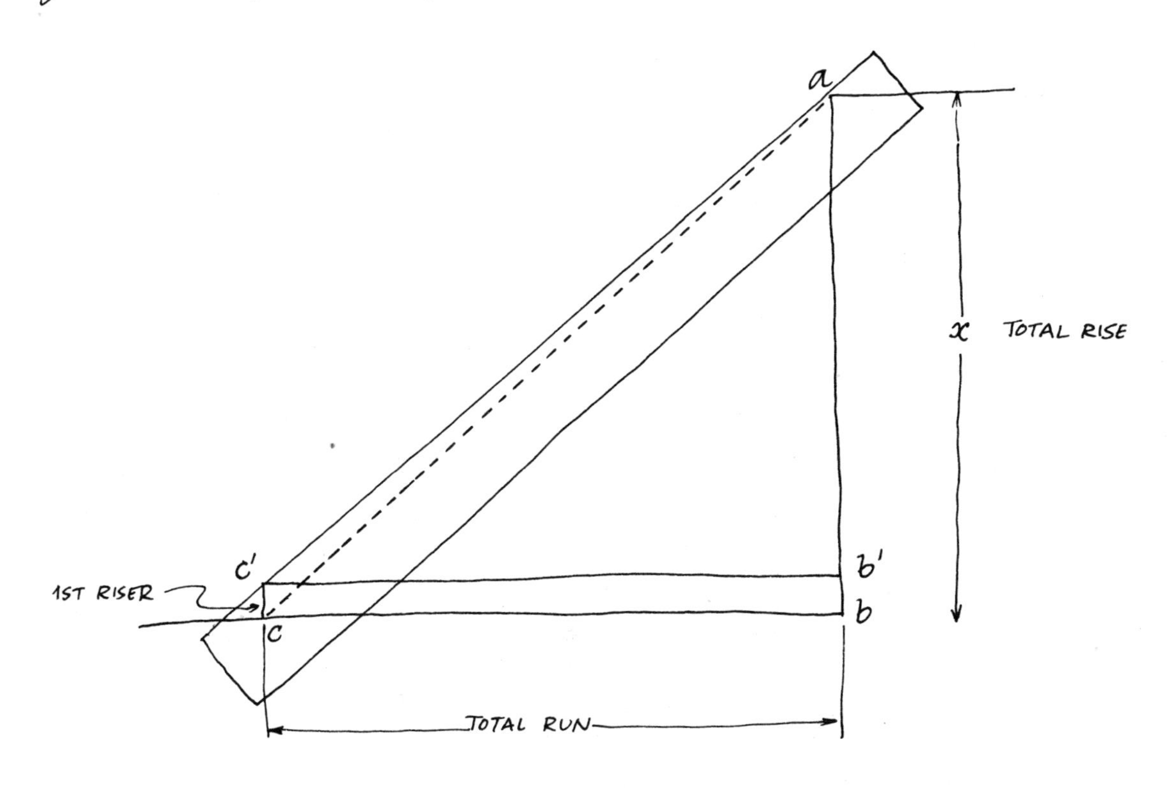

Transfer a b and a c from your previous working drawing (total rise and total run) and connect a c.

Erect a perpendicular at c to c', to equal the height of a riser. Now connect ac' which is the "step length" of the stringer. Draw c'b' parallel to cb.

In the triangle ab'c', $ac' = \sqrt{ab'^2 + b'c'^2}$

This is the length which must be stepped off by the dividers into as many steps as there are.

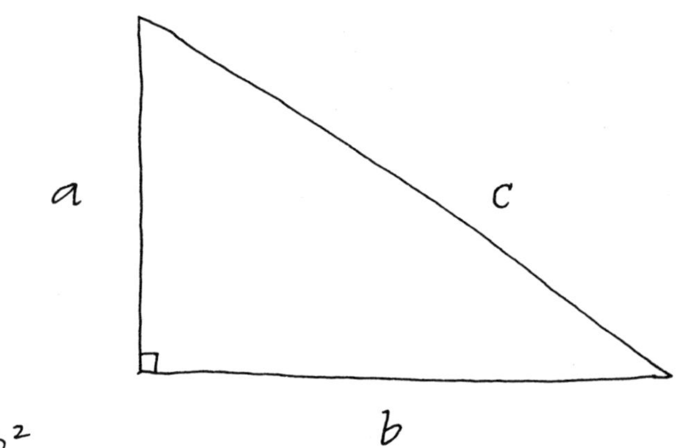

$c^2 = a^2 + b^2$

PYTHAGORAS' THEOREM

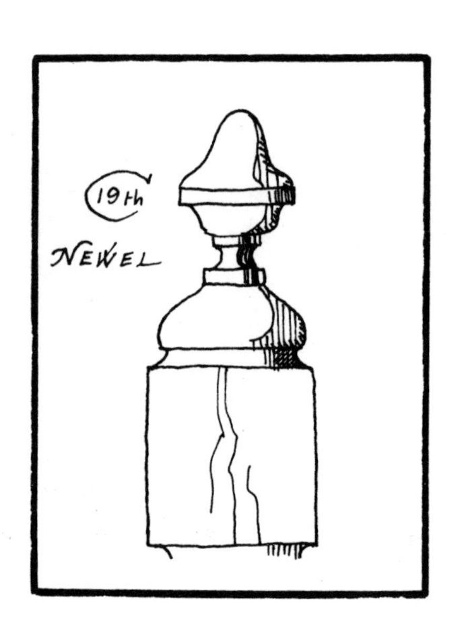

Depending on the wood used and its dimensions, stairs over a certain width may require a center stringer for extra support. If the outside stringers are open stringers, then you may use the waste portions to build up the center stringer as shown below.

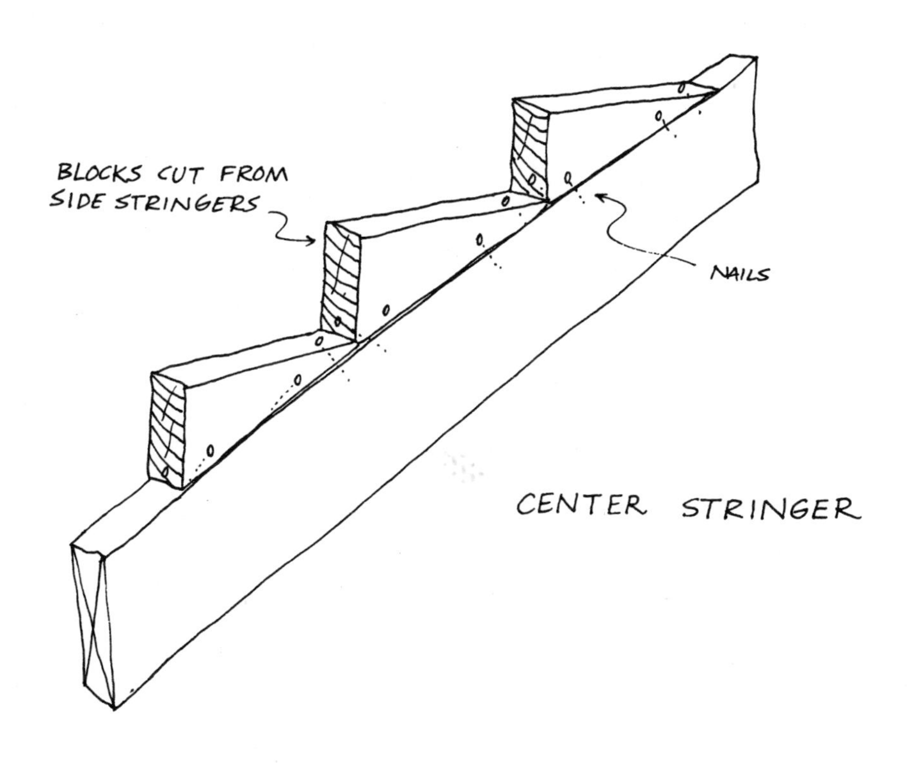

The best material is usually oak, since this is very hard wearing. At least full inch should be used, but five-quarter (31.5 mm) is better. If the stair is not to get much wear, such as a little-used basement stair, then of course some less expensive wood may be employed.

NEWELS

The only important requirement for the newel is that it be fixed very securely, since it holds the foot of the stringer and the handrail.

The actual design is capable of almost infinite variation, providing a square section to receive the end of the stringer and a square section to receive the end of the handrail is provided.

The stringer should be mortised into the newel as shown.

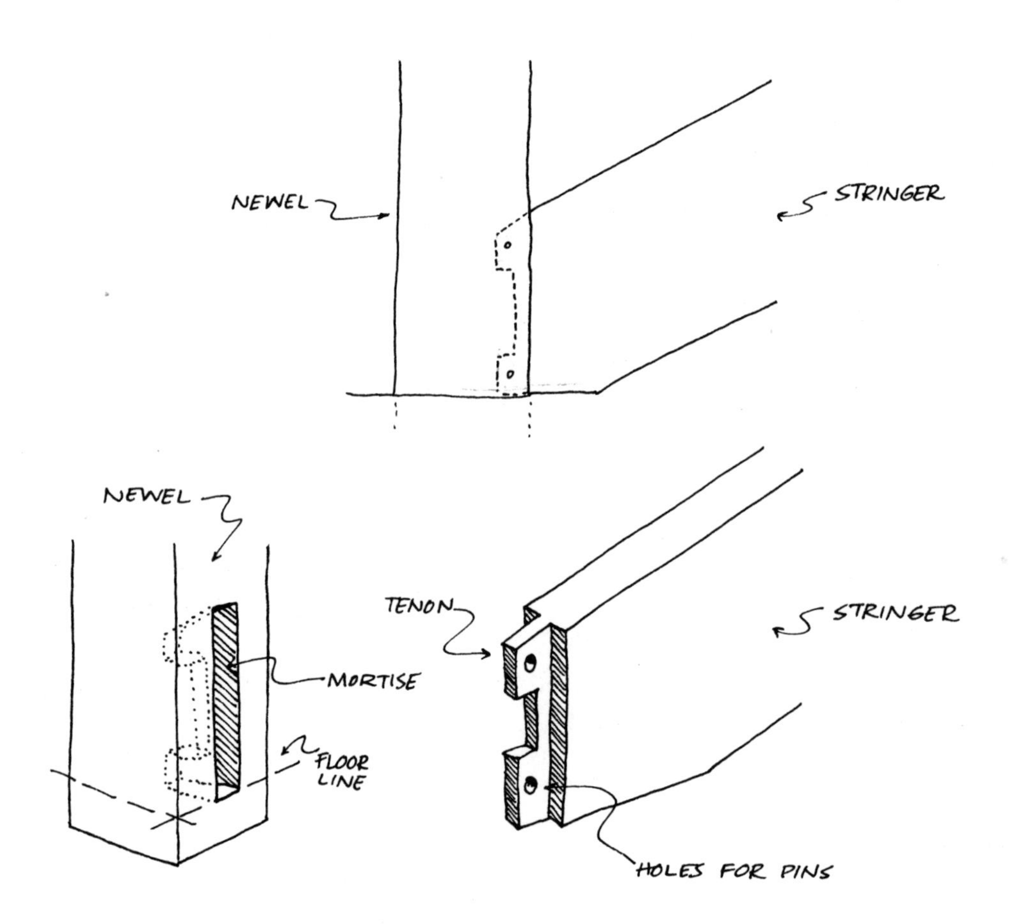

Securing the newel to the floor is all-important. Ideally, the newel should run down beneath the floor and be spiked to the side of a joist.

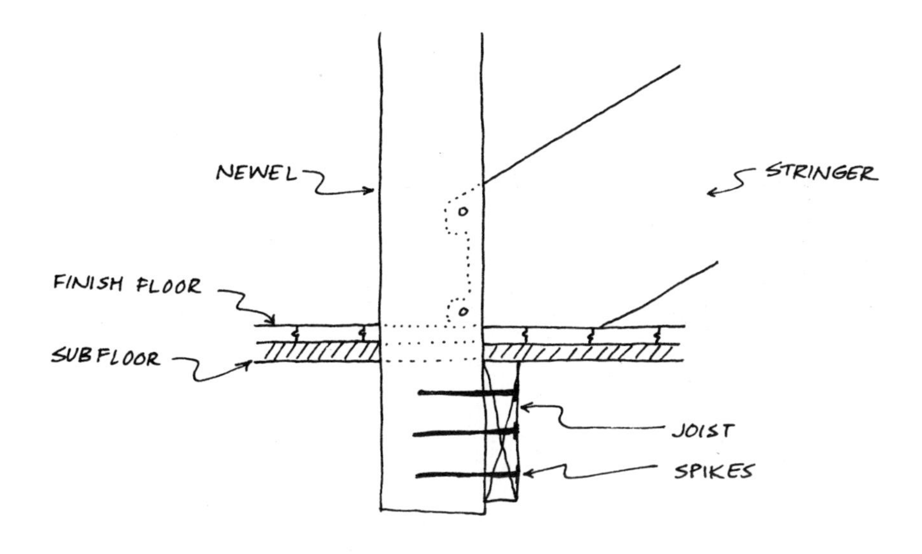

If this is not possible, try furring the joist out, or even cutting part of the newel away.

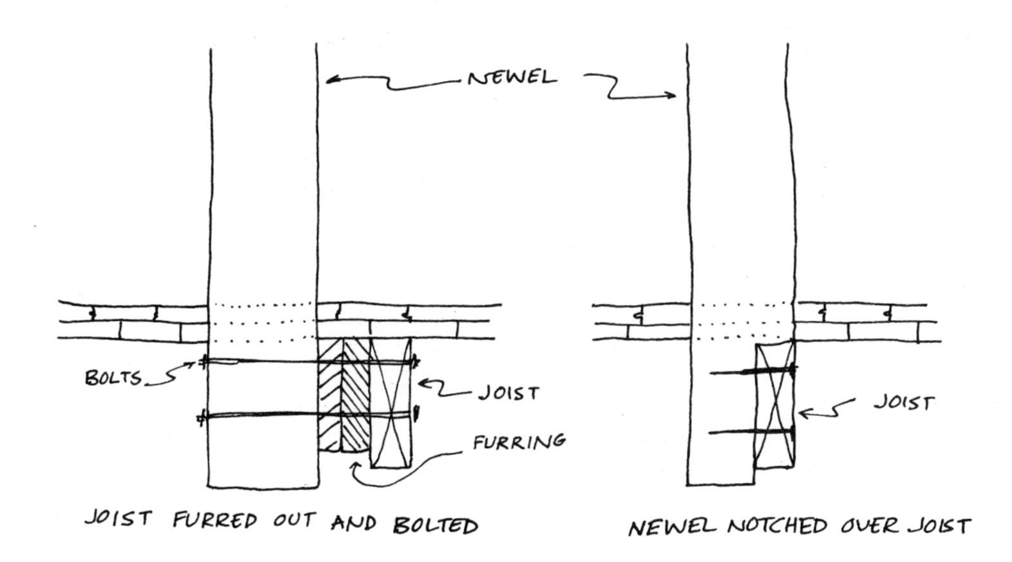

JOIST FURRED OUT AND BOLTED

NEWEL NOTCHED OVER JOIST

If the joists run the wrong way, or are too far away from the foot of the newel, make a pair of bridges as shown.

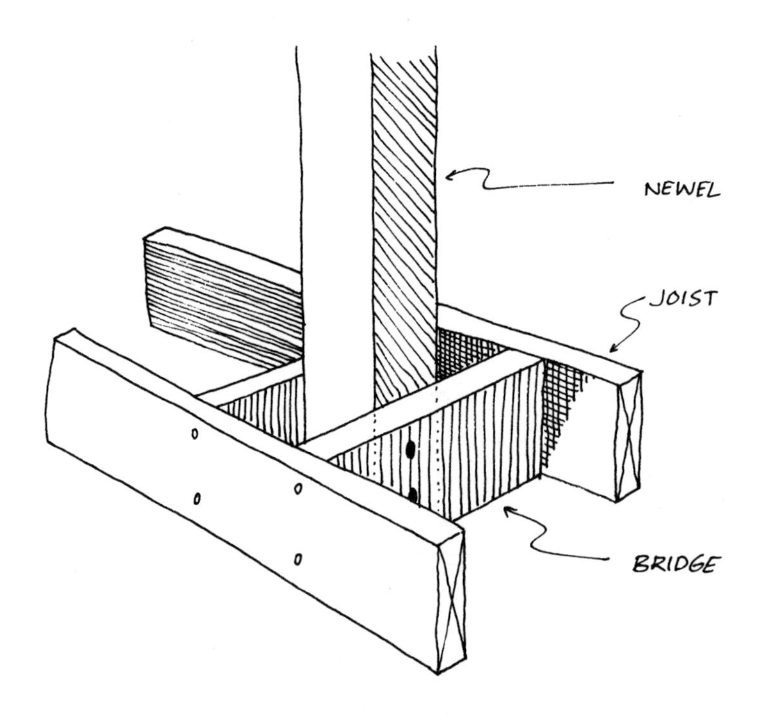

If none of the above is possible, bolt the newel to the floor, using a reinforcing block and mortising into the finish floor.

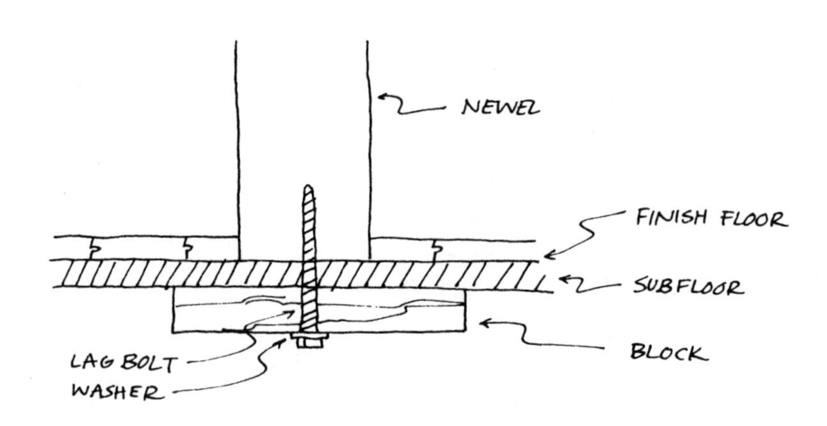

HANDRAILS

It is no exaggeration to say that as much could be written on the complexities of handrails as has been written in this entire book. But labor is more expensive today than it was formerly and the average handrail has become a far more straightforward affair.

It may be bought in lengths at any good lumberyard, often in a selection of mouldings, together with the appropriate hardware (or ironmongery as it is called in Britain).

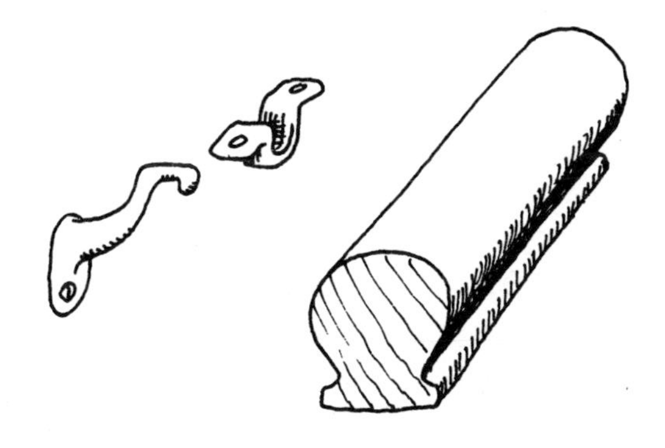

As stated in the introduction, this book is not meant to be an exhaustive treatise on all aspects of joinery, but a practical guide for the layman or beginner building a simple house.

However, if you have come this far, you will have a good foundation for more advanced joinery, should you so desire. There is a lot more to the subject, probably too much for any one person. As mentioned earlier, there are craftsmen totally occupied with but single aspects of this book. And, unless we all return to caves and trees, there will always be a need for this knowledge, from its simplest form to its most complicated.

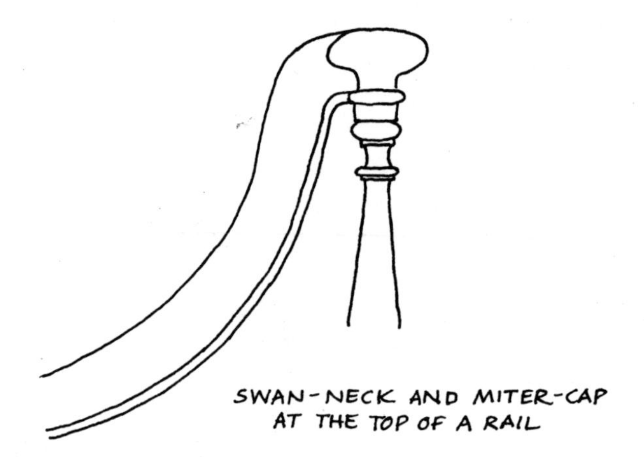

SWAN-NECK AND MITER-CAP
AT THE TOP OF A RAIL

Index

Finis